Landscape Architecture - Processes and Practices Towards Sustainable Development

Edited by Luís Loures and Mustafa Ergen

Published in London, United Kingdom

IntechOpen

Supporting open minds since 2005

Landscape Architecture - Processes and Practices Towards Sustainable Development
http://dx.doi.org/10.5772/intechopen.87654
Edited by Luís Loures and Mustafa Ergen

Contributors
Adel Zeggaf Tahiri, G. Carmi, M. Ünlü, Grzegorz Szewczyk, Krzysztof Lipka, Piotr Wężyk, Karolina Zięba-Kulawik, Monika Winczek, Jon Burley, Yoichi Kunii, Luis Loures, Paige O'Keefe, Marifaye Regina Villanueva, Rui Alexandre Castanho, José Cabezas Fernández, José Manuel Naranjo Gómez, José Martín Gallardo, Luis Fernández-Pozo, Sema Yilmaz Genç, Sérgio Lousada, Cláudia Fernandes, Giulia Olivetti, Krishna Rani Barai, Funmilayo M. Adedire, Moshe Gophen, Mustafa Ergen

First published in London, United Kingdom, 2021 by IntechOpen
IntechOpen is the global imprint of INTECHOPEN LIMITED, registered in England and Wales, registration number: 11086078, 5 Princes Gate Court, London, SW7 2QJ, United Kingdom
Printed in Croatia

British Library Cataloguing-in-Publication Data
A catalogue record for this book is available from the British Library

Additional hard and PDF copies can be obtained from orders@intechopen.com

Landscape Architecture - Processes and Practices Towards Sustainable Development
Edited by Luís Loures and Mustafa Ergen
p. cm.
Print ISBN 978-1-83968-376-3
Online ISBN 978-1-83968-377-0
eBook (PDF) ISBN 978-1-83968-378-7

For EU product safety concerns:
IN TECH d.o.o., Prolaz Marije Krucifikse Kozulić 3, 51000 Rijeka, Croatia, info@intechopen.com or visit our website at intechopen.com.

Meet the editors

Luís Loures, Ph.D., is a landscape architect and agronomic engineer, and vice president of the Polytechnic Institute of Portalegre, Portugal. He has a Ph.D. in Planning and a postdoctorate degree in Agronomy. He is also a researcher at VALORIZA – Research Centre for Endogenous Resource Valorization, Polytechnic Institute of Portalegre, and CinTurs – Research Centre for Tourism, Sustainability, and Well-being. At the latter, Dr. Loures is a researcher on several financed research projects focusing on urban planning, landscape reclamation, and urban redevelopment, and the use of urban planning as a tool for achieving sustainable development.

Mustafa Ergen graduated from the Department of Landscape Architecture, Abant İzzet Baysal University, Turkey, in 2000. In 2005, he completed his first master's degree in Urban and Regional Planning at the Gebze Institute of Technology, Turkey, and his second master's degree in Landscape Architecture at Anhalt University of Applied Sciences, Germany, in 2006. During 2007–2008, he studied Geographic Information Systems and Remote Sensing at the Mediterranean Agronomic Institute of Chania, Greece, and was granted a specialization diploma in Environmental Management. He also received a Dr. Eng. from the Technical University of Dortmund, Germany. Currently, he works in the Department of Architecture, Siirt University, Turkey.

Contents

Preface

As generally acknowledged, landscape architecture extends across numerous subject areas from heuristics, aesthetics, and sciences to statistical decision-making, technology, and arts. Landscape architecture ranges from small projects and details to global ecology and ecosystems. For this reason, the profession incorporates information from the social sciences, natural science, natural history, ecology, and the medical sciences. Often a good landscape architect is someone who is interested in many subjects and is able to integrate them thoughtfully. In this regard, landscape architecture, generally described as a multi-disciplinary field that incorporates several branches of knowledge, is perceived as a discipline able to promote the definition of land uses, thus enabling landscape alterations that might promote sustainable development, protect the environment, preserve natural and cultural assets, and improve people's quality of life.

In fact, crossing a wide range of research domains and issues associated with landscape planning, design, and redevelopment, this book covers a large number of topics related to landscape architecture, assessing the impact of contemporary needs and constraints and landscape management strategies on planning, ecosystems, and landscape design. The book presents specific case studies representing a vast array of landscape planning and design projects that go beyond conventional practice, including not only a diverse amount of end uses but also multidisciplinary methodological approaches.

As a landscape architect who then became an agronomic engineer with research interests deeply rooted in the field of sustainability, landscape planning, and development, the main goals of most of my research are directly connected to fitting landscape architecture as a tool to promote and assure that planning and design processes meet the needs and desires of contemporary life. All this, bearing in mind the comprehensive view of the different components of landscape architecture, acknowledging the need for an interrelated analysis of the ecologic, cultural, and socioeconomic issues in each and every single planning and design process.

Luís Loures
VALORIZA – Polytechnic Institute of Portalegre,
Portalegre, Portugal

Mustafa Ergen
Faculty of Fine Art and Design,
Department of Architecture,
Siirt University,
Siirt, Turkey

Section 1

Introduction

Chapter 1

Introductory Chapter: Landscape Architecture - The Gatekeeper of Sustainable Development

Luís Loures

1. Introduction

The connection between man and the natural world is no more that of a harmonious accord, but it is often a relation of contrast which has caused and which still causes two big problems that sometimes look not being solvable: the problem of the environment pollution and the problem of the progressive exhaustion of the resources, thus imposing to the community high social costs [1–3]. This results in a new attitude towards the environment, which is not a utopic return to the past but, instead, the attitude of identifying the meaning of a sustainable development as a process of change, in which resource utilization, investment, technological development, and institutional changes are in a reciprocal harmony, increasing current and future potential of satisfying human needs and aspirations [4–6]. This repurposes a renewed connection man-nature, according to an ecosystem approach, meant to promote the identification, protection and enhancement of natural resources, in order to reinstate the necessary equilibrium between the built and the natural environment.

This necessary equilibrium, coupled with the impossibility of an unlimited growth in a world of limited resources and the extensive diffusion of sustainability as a crucial conceptual approach towards development, constitute a distinctive mark that led countries to operate sustainable development as one of the main planning paradigms of the last decades. Thus, landscape must be understood as a global entity, which have several components, that must be analyzed, managed and protected in order to achieve sustainable development though the application of specific planning and design strategies based on landscape architecture principles. Still, regardless of the conceptual approach used, the fact that the landscape constitutes a dynamic structure in constant transformation, highlights the fact that any intervention aiming at its management, planning and/or alteration must be based on a deep knowledge of its characteristics, history, structure, and services.

This reality is increasingly acknowledged, in a society in which, production and consumption patterns have become completely unsustainable, accelerating the exhaustion of natural resources, fostering environmental pollution and promoting landscape loss, thus hastening problems that impose high social costs to citizens and communities all over the world. It is therefore necessary to take up a new attitude towards landscape and the environment, by using and implementing sustainable design principles and approaches, anchored in landscape architecture as gatekeeper of sustainable development, therefore promoting the coordinates for environmental management and protection, social equity and economic prosperity, as the base for landscape sustainability.

In this regard, considering that landscape is the visible expression of the physical and biological components and of human activities and settlements in a given territory, immediately reflecting changes in their occupation and use (**Figures 1–3**), and that, for this reason, it is continuously changing as a result of multifaceted and cooperating natural processes coupled with planned and unplanned actions by man [7, 8], global concerns raised by landscape transformation processes worldwide, highlight the need to rethink landscape while protecting the environment. This is evident in each and every single landscape, but, especially true for previously developed ones which are currently derelict and abandoned [9], posing different but complementary challenges not only to the society but also to landscape architecture as discipline.

In fact, the relevance and popularity of landscape architecture approaches and projects in an increasingly changing environment are progressively recognized as

Figure 1.
Natural intermittent river – Caia, Portalegre. Photo credits – Luís Loures –

Figure 2.
Agricultural landscape – Elvas, Portugal. Photo credits – Luís Loures –

Figure 3.
Artificial lake – Polytechnic Institute of Portalegre – Campus, Portugal. Photo credits – Nuno Bilé – Used by permission, all rights reserved.

an adequate discipline to answer specific development questions such as: Which are the crucial ecological functions that any project should account for? What makes landscapes valuable and appealing? What constitutes good design? What is the contribution of landscape architecture to sustainability? What makes landscapes underutilized? What obstacles keep derelict and abandoned landscapes from being recovered or transformed? Who should be responsible for landscape planning and management both at local, regional and national levels? Still, new methodologies and frameworks are needed, in a period when "*(...) that seemingly old-fashioned term landscape has curiously come back to vogue*" [9–11], it is urgent to reinvent the way in which landscapes are planned, managed, recovered or transformed, considering sustainability as a basis for development in which environmental issues, historic and cultural values, economic opportunities, and social needs are considered at the same level.

2. Landscape architecture – Integrating knowledge towards sustainable development

Considering that landscape architecture, landscape planning, landscape reclamation, landscape management and landscape preservation are part of the same process [1–3], existing knowledge on the biophysical character of a landscape, and on the way this character restricts or favors the implantation and the development of human activities, is essential to achieve its sustainability. This idea is also valid to the cultural and esthetic characters, because history, previous land use, people's culture, and landscape scenic and biological values, constitute essential factors for the acceptance of new landscapes by the population.

This idea explains how important it is to integrate all the landscape components, not only in planning and management activities, but in any kind of project concerned with the creation or change of the public space, because one may never forget that the main idea of any kind of project, must be "public space is for the public", and that's the reason why their needs, wishes and aspirations are one of the most important aspects in landscape architecture project and planning activities [12–14].

Anyway, landscape planning, design, management and/or redevelopment activities, provide constant new opportunities for those who have the desire and the ability to seize landscape, regardless of its nature [15–18], for this reason landscape architecture is considered to be, a significant resource for achieving sustainable development contributing as well to improve life's quality. In this regard, landscape design processes need to be thought in terms of sustainability and/or sustainable development, "*terms that get used a lot these days, and which since their appearance have been faced as new development paradigms introduced in land-use matters, merging social, economic and environmental dimensions*" [9], pushing regions and nations to work together in the establishment of new principles, frameworks and methodologies towards sustainable development.

The present book addresses the importance of the environmental, cultural, esthetic and economic contributions in landscape analysis, planning and design, while creating the common ground for enabling a broad understanding of the relationship established between landscapes, territories and human occupation along time. Consequently, aiming at the total understanding of landscape architecture domains, the envisioned book adopts a holistic vision of landscape, integrating its various components: the environmental, which includes both physical and biological parts of the ecosystems; the cultural, where both historical factors, genius loci, identity, sense of belonging, and the narrative qualities of the landscape are considered; the socio-economic, referring not only to the social factors and economic activities determining the human action permanently constructing and changing the landscape, but also the regulations and legal instruments which affect these activities, approaching the contributions that the each component has in landscape design, planning, management and/or redevelopment.

Author details

Luís Loures
VALORIZA—Research Center for Endogenous Resource Valorization—Polytechnic Institute of Portalegre, Portugal

*Address all correspondence to: lcloures@ipportalegre.pt

References

[1] Loures, L. & Panagopoulos, T., 2010. Reclamation of derelict industrial land in Portugal - greening is not enough. International Journal of Sustainable Development & Planning, Vol. 5(4) 343-350.

[2] Loures, L., Loures, A., Nunes, J. e Panagopoulos, T., 2015. Landscape Valuation of Environmental Amenities throughout the Application of Direct and Indirect Methods. Sustainability 7(1), 794-810.

[3] Loures L. e Panagopoulos T., 2007. Sustainable reclamation of industrial areas in urban landscapes. Em: Kungolas, A, Brebbia, C. e Beriatos, E. (Eds.), Sustainable Development and Planning III, WIT Press, Southampton, pp. 791-800. ISBN:978-1-84564-102-3.

[4] Brueckner J. Urban sprawl: Diagnosis and remedies. International Regional Science Review. 2000;23(2):160-171

[5] Johnson M. Environmental impacts of urban sprawl: A survey of the literature and proposed research agenda. Environment and Planning A. 2001;33(4):717-735

[6] Reed P, editor. Groundwell: Constructing the Contemporary Landscape. New York: The Museum of Modern Art; 2005

[7] Ferreira, V., Barreira, A., Loures, L., Antunes, D. & Panagopoulos, T., 2020. Stakeholders' Engagement on Nature-Based Solutions: A Systematic Literature Review. Sustainability, 12 (2), pp 640.

[8] Kurowska-Pysz, J., Castanho, R. & Loures, L., 2018. Sustainable Planning of Cross-Border Cooperation: A Strategy for Alliances in Border Cities. Sustainability, 10 (5), pp 1-26.

[9] Loures, L., 2019. Introductory chapter: land-use planning and land-use change as catalysts of sustainable development. IN: Loures, L. (Ed.), Land Use - Assessing the Past, Envisioning the Future. IntechOpen. ISBN: 978-1-78985-704-7.

[10] Corner J. 2006. Terra Fluxos. In: Waldheim C, editor. The Landscape Urbanism Reader. New York: Princeton Architectural Press.

[11] Waldheim C, eds, 2006. The Landscape Urbanism Reader. New York: Princeton Architectural Press.

[12] Loures, L. & Crawford, P., 2008. Democracy in progress: using public participation in post-industrial landscape (re)-development. WSEAS Transactions on Environment and Development, Vol. 4(9) 794-803.

[13] Loures L, Santos R, Panagopoulos T., 2007. Urban parks and sustainable city planning—The case of Portimão, Portugal. WSEAS Transactions on Environment and Development. 2007;3(10):171-180

[14] Meireles, F. & Loures, L., 2017. Public Participation Project-based Learning in Landscape Architecture. WSEAS Transactions on Environment and Development, Vol. 13:441-451. ISSN: 2224-3496.

[15] Loures, L. & Panagopoulos, T., 2010. Reclamation of derelict industrial land in Portugal - greening is not enough. International Journal of Sustainable Development & Planning, Vol. 5(4) 343-350.

[16] Ferreira V, Panagopoulos T, Andrade R, Guerrero C, Loures L., 2015. Spatial variability of soil properties and soil erodibility in the Alqueva dam watershed, Portugal. Solid Earth;7(301-327)

[17] Loures L, Loures A, Nunes J, Panagopoulos T., 2015 Landscape

valuation of environmental amenities throughout the application of direct and indirect methods. Sustainability. 7(1):794-810

[18] Lu D, Burley J, Crawford P, Schutzki R, Loures L., 2011. Quantitative methods in environmental and visual quality mapping and assessment: A Muskegon, Michigan watershed case study with urban planning implications. In: Urban Planning. IntechOpen; pp. 127-142

Section 2

Landscape Change: Uses and Services

Chapter 2

Land Uses Allocation: The Execution of an Artificial Beach and Its Complementary Infrastructures – Madeira Island – Machico, Portugal

Sérgio Lousada, Luís Loures and Rui Alexandre Castanho

Abstract

The present study aims to propose the creation of an artificial beach in the municipality of Machico and its complementary infrastructures, located on the south-eastern coast of the Madeira Island (Portugal). Machico's beach sand consists of a mixture of black volcanic sand and round basalt stones. Usually, it has clear waters and a quiet sea. This beach also has a mooring infrastructure, thus allowing access to the sea. In order to achieve this study's main goals, it was initially carried out an extensive review and bibliographic research. Subsequently, a sand beach and its shelter groins were simulated and designed to hypothetically promote the retention of the sand and mitigate the tidal effects. In addition to model the beach dynamics, an extensive characterization of the extreme maritime regime was performed. The model was developed based on topographic and hydrographic site surveys and the data using Wave Watch III model at 32.0°N, 17.0°W—obtained from SONEL web page, which gathers new data every 6 hours. Besides, the study also contemplated the analysis of the sea inundation quota for a return period of 100 and 500 years and its development along with the Master Plan of the City of Machico. Furthermore, some final remarks and conclusions will be shown; besides, some future projects should be developed to expand the knowledge of this thematic field.

Keywords: artificial beach, beach dynamics, coastal protection, extreme scalar wave height, GIS, maritime climate, maritime flood level, return period, sediment transport, waves

1. Introduction

Currently, there is enormous progress pressure in the coastal zone worldwide, which has resulted in the planning or construction of a large number of coastal developments. This situation requires more than protecting the existing coasts against natural risks, such as coastal erosion and coastal flooding. Typically, there is a need to rehabilitate many coastal areas under pressure from land development in the past and coastal erosion and degradation. However, in some regions, the high demand is not satisfied only with the rehabilitation of existing beaches [1].

In RAM (Autonomous Region of Madeira), the large concentration of hotels along the south coast, resulted in the reduction of free access spaces for the population to the sea, a situation that was compensated by the construction of several bathing complexes, cases of Ponta Gorda, Poças do Governador, Doca do Cavacas, Barreirinha, Ponta Delgada, among others [2].

From the rolled pebbles to the black sand, to the artificial and the various bathing complexes with natural pools, it is possible to find beaches for all tastes in Madeira; in fact, this created a competition with Porto Santo, until recently the only island in the archipelago with its nine-kilometer-long fine yellow sand beach [2].

In Madeira, initially, the beaches were all of rolled pebbles, with only one of black sand, the 'Prainha' in Caniçal (**Figure 1**), at the eastern end of the island, which is difficult to access, and some private and public bathing complexes in Funchal [2].

But about a decade ago the scenario started to change and one of the innovations was the import of sand from North Africa to build the artificial beach in Calheta (**Figure 2**) [2].

Calheta was the first beach in Madeira to import sand, having been inaugurated in 2004 and with two slope breakwaters as a form of protection. Praia da Calheta is a beach located in the parish of Calheta, on the island of Madeira, in Portugal, with a length of 100 meters. It is often sought after by canoeists and windsurfers [3]. This project was so successful that it was copied in the municipality of Machico, in the extreme east [2]. Located on the right bank of the mouth of Ribeira de Machico, Praia da Banda d'Além (**Figure 3**) is a beach that allows an immediate dive for those in the center of Machico [4].

With about 70 meters in length, the beach has locker rooms, changing rooms, showers, bathrooms, parking, and guarded during the bathing season [4]. This yellow sand beach is one of the few references of its kind on Madeira Island. It is

Figure 1.
"Prainha", Machico (source: https://www.madeiraallyear.com).

Figure 2.
Calheta artificial beach (source: https://www.madeiraallyear.com).

Figure 3.
Banda d'Além beach, Machico (source: https://www.madeiraallyear.com).

Figure 4.
Machico Bay (source: Authors).

framed in the emblematic Machico bay (**Figure 4**) inserted in a pleasant environment, making it, therefore, one of the favorite places regionally [4].

The introduction of sand from North Africa on the beaches of Madeira in recent years changed the appearance of the island but also extended the tourist offer for those seeking sun and sea [2].

The proposal presented is intended to create an artificial beach in the municipality of Machico and the necessary infrastructures, continuing the success of Praia da Banda d'Além, expanding the offer, and contributing to the economic and social development of the municipality and, consequently, RAM.

In addition to model the beach dynamics, an extensive characterization of the extreme maritime regime was performed in order to portray the analysis between the "MFL" maritime flood level, designates the sea level reached in exceptional situations and its iteration with the PDMM – essential instrument for spatial planning in the municipality of Machico. This part of the work has been previously developed, for example, in Colombia, where methodologies were proposed to estimate the long-term Maritime Flood Level (MFL) on a regional scale on the Caribbean coast and on the Pacific coast where a study was carried out in which a study was carried out. In fact, measures of vulnerability and adaptation to flooding in coastal and insular areas of Colombia were analyzed, depending on the characteristics of the population and its infrastructure [5].

2. Workflow: the used methods

In this study, a quantitative methodology is adopted, which is characterized by the use of quantification, data collection process, and treatment of these data through statistical techniques. It is often applied to sciences in descriptive studies that seek to discover and classify the relationship between variables, ideal for the physical–mathematical study.

Firstly, an essentially theoretical collection of information to be used in this study is carried out. So, it was established which support software is needed and which physical parameters contribute with greater significance in executing a project of this magnitude.

Based on the selected drawing and spreadsheet software and making use of the collected, organized, and treated data, a project consisting of a descriptive memory, a set of construction plans, a construction contract documents/special technical conditions, the respective health and safety budget and plan, were prepared and developed.

After elaborating on the project, observations and analyses are carried out on the set of written and drawn pieces and their importance during the design, execution, and exploration phases.

At the end of the study, the conclusions and intrinsic recommendations are presented, being this work based on the following organization chart (**Figure 5**).

Figure 5.
Organigram (source: Authors).

3. Case study

The Madeira Archipelago, an integral part of Portuguese territory, is located in the Atlantic Ocean 978 km southwest of Lisbon. Of volcanic origin, it is formed by the islands of Madeira (736 km^2), Porto Santo (43 km^2), Desertas (14 km^2) and Selvagens (4 km^2). Only the first two islands are inhabited, making up the other natural reserves [6, 7].

The island of Madeira has a very rugged orography (**Figure 6**), with the highest points being Pico Ruivo (1,862 m) and Pico do Areeiro (1,818 m). The relief, as well as the exposure to the prevailing winds, means that there are several microclimates on the island, which, together with the exotic nature of the vegetation, constitutes an important factor of attraction for tourism, the main activity in the region [6, 7].

There are no major thermal variations throughout the year, keeping the climate mild with average temperatures around 22°C (maximum) and 16°C (minimum) [6, 7].

The sea water temperature, due to the influence of the hot Gulf current, remains at 22°C in the summer, gradually cooling until reaching 17°C at the end of winter [6, 7].

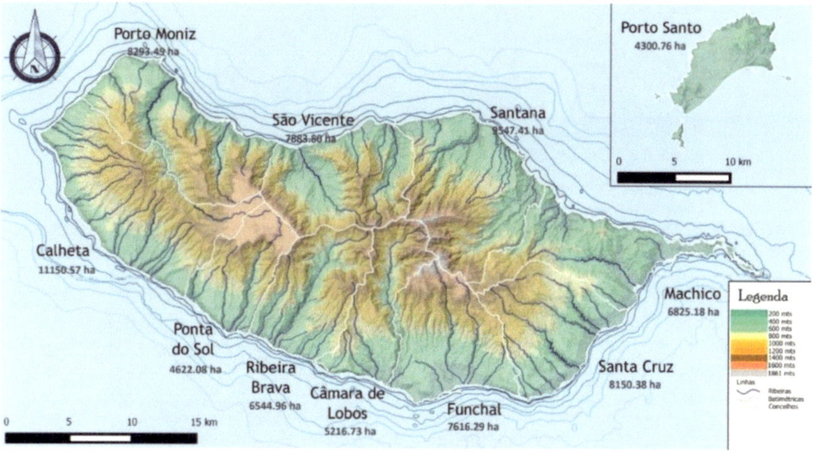

Figure 6.
Geomorphology of Madeira Island and Porto Santo), main and bathymetric streams with the limits and official areas of the municipalities (Author: L. C. Antunes).

Figure 7.
Proposal for a new beach and associated infrastructures (source: Authors).

Figure 8.
Place to intervene (source: Authors).

The case study is located on the south-eastern coast, in the municipality of Machico (Madeira Island) with the following geographic coordinates: Latitude: 32°42′58.15″N, Longitude: 16°45′51.90″W (**Figures** 7 and **8**).

Figure 9.
Beach detail (source: Authors).

Machico beach (South) is a beach with a mixture of black sand and basalt stones (**Figure 9**). It has clear waters and a calm sea with a mooring (pontoon) structure that allows access to the sea or coast. The coast has several catering services and similar others.

The aim is to simulate the creation of a sand beach on the coast of Machico (South), simultaneously with the dimensioning of shelter groynes to promote sand retention and mitigate the undulation on the site (hypothetical project). One possibility for creating this beach is to use the dredged sand in the port of Funchal. For this, it is necessary to make a comparison between the profile of Garau and the real profile of the beach, measured in situ.

Additionally, there is a structural rehabilitation and expansion of the mooring structure (pontoon). Finally, the MFL is also calculated for a return period of 100 and 500 years. In order to model and design this analysis, a characterization of the extreme maritime regime is needed [8, 9].

As starting data, the local topographic and hydrographic surveys is provided, as well as the wave data (Wave Watch III model at 32.0°N17.0°W, obtained at www.sonel.org. – the historical series is wider (1952 to 2012), with data every 6 hours), as presented in **Figure 10**.

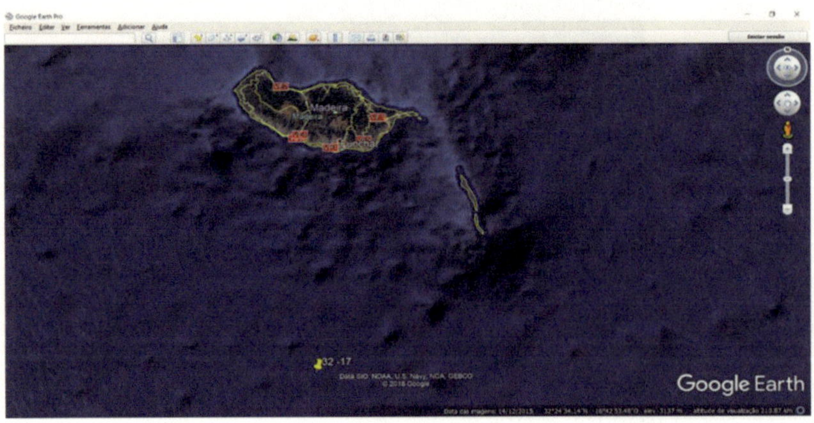

Figure 10.
Location of the buoy in relation to the coast (source: Author).

4. The processes to assess the maritime climate

Prerequisites:

- Choose a point on the coast where it is assumed that the Maritime Work will be projected, for which the Maritime Climate in Deep Waters will be calculated;

- Choose the type of Maritime Work to be projected, in order to calculate the Lifetime (L) and the Risk (R) that represents the calculation time.

The seven steps indicated in **Figure 11** were followed to obtain the Maritime Climate in Deep Waters at the chosen coast point.

The chosen coast point is located in the municipality of Machico – Machico Beach (South) with coordinates 32°42′59.23″N; 16°45′52.02″W. **Figure 12** shows the area where the project will be developed (hypothetical maritime work), in this case, an artificial beach, on which the study will be carried out.

The information used in the development of the maritime climate was SONEL database (database GOW – IH Cantabria de Santander – www.sonel.org), "Waves" tab. The point on the coast is approximately 82.5 km from the selected buoy located in the south-west of the island of Madeira (32°00′00.00″N; –17°00′00.00″W).

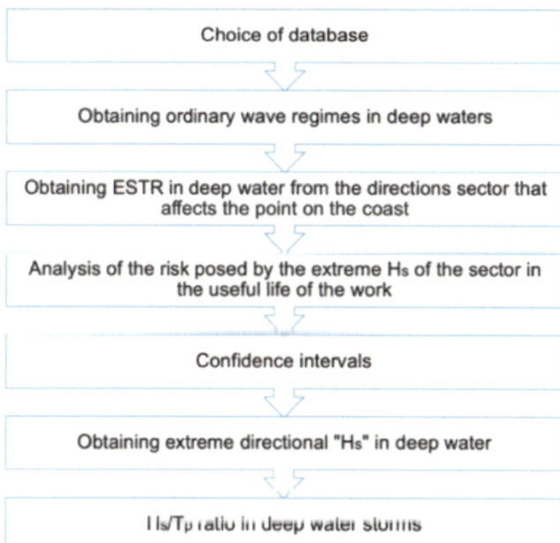

Figure 11.
Calculation methodology (source: Authors).

Figure 12.
Proposal for a new beach and associated infrastructures (source: Authors).

Figure 13.
Range of directions that affect the point on the coast chosen (source: Authors).

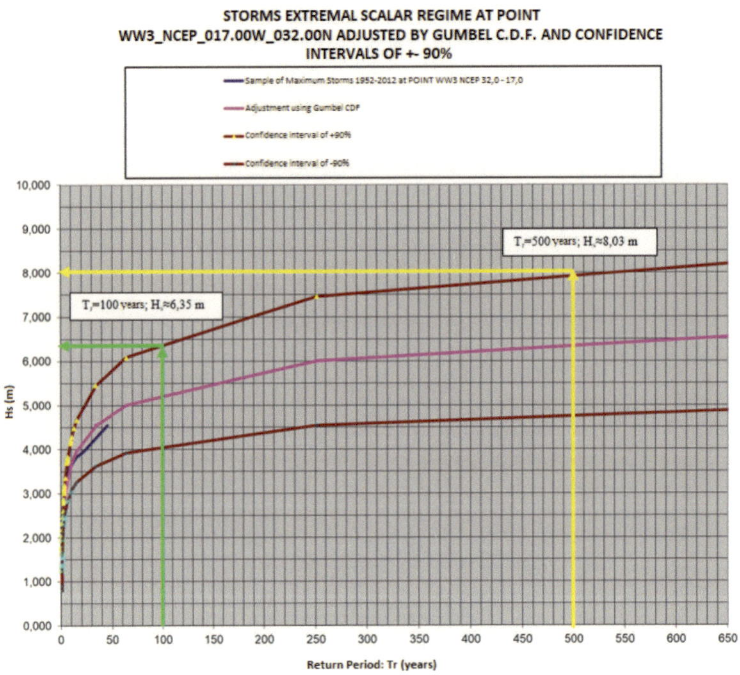

Figure 14.
Extreme scalar temporal regime adjusted by the C.D.F. Gumbel together with the confidence bands for the ± 90% percentile (source: Authors).

In this specific case, and for the point on the coast chosen, the directions to be considered include the sectors SSW-S-SSE-SE-ESE of wind rose, as shown in **Figure 13**.

In order to obtain the Extreme Scalar Temporal Regime (ESTR), it is first necessary to obtain the maximum annual directional significant wave height (hs, max) and the peak period (tp – seconds).

Figure 14 shows the graph on the Cartesian axes – Return Period Selection (Tr) vs. extreme scalar wave height (Hs) – for our 59-year-old sample (1952–2011) (chosen database, SONEL (www.sonel.org), "Waves" separator, namely the buoy located in the south-west of the island of Madeira (32°00′00.00″N; –17°00′00.00″W)), through the C.D.F. Gumbel together with the confidence bands for the ± 90% percentile.

After executing the previous task, we have the following:

- C.D.F. Gumbel (asymptote I) we have an extreme scalar Hs for the confidence band, percentile + 90% whose value is Hsee = 6.35 m, for Tr = 100 years;

- C.D.F. Gumbel (asymptote I) we have an extreme scalar Hs for the confidence band, percentile + 90% whose value is Hsee = 8.03 m, for Tr = 500 years.

Then focus on the calculation of the maritime flood level of the beach of Machico (South) – municipality of Machico, on the south coast of Madeira (Portugal). Along with that developed over the previous information, it is intended to obtain a comprehensive view of the local dynamics in order to schematize its performance.

The MFL, designates the sea level reached in exceptional situations. It depends on the following factors: characteristics of the swell or the incident storm (Hs, Tp), slope of the beach and the existence or not of coastal defenses.

It is possible to distinguish several levels of maritime floods depending on their origin: the simple maritime, the simple rainwater and the rain-sea combination.

A comparative analysis in two dimensions of the area and topographic levels covered by the MFL was carried out for the beach of Machico (South) – municipality of Machico (Portugal), as well as its iteration with the Machico Master Plan – basic instrument for spatial planning in the municipality of Machico. That said, with the use of Microsoft Excel and AutoCad software, it was possible to develop this analysis. Thus, the order described below allows us a better understanding of the above-mentioned:

- Hsedir for the range of directions that affect the work, which multiplied by the directionality coefficient, Cd, of the respective direction, gives the maximum value of Hsee = 3.42 m;

- C.D.F. Gumbel (asymptote I) we have an extreme scalar Hs for the confidence band, percentile + 90% whose value is Hsee = 6.35 m, for Tr = 100 years;

- C.D.F. Gumbel (asymptote I) we have an extreme scalar Hs for the confidence band, percentile + 90% whose value is Hsee = 8.03 m, for Tr = 500 years.

Regarding to the Machico Master Plan and its resolution, we will be sent to:

- Article 29 – Identification of spaces – Depending on the existing or proposed dominant use, the following classes and subclasses of spaces are considered, identified in the planning plan:

    a. Natural spaces – Natural spaces for recreational use (beaches);

- Article 31 – Identification of the operational planning and management units (UOPG) – Without prejudice to the elaboration of municipal plans of a lower hierarchy for the entire area of urban land production in the municipality, the UOPG identified in this Plan and which are considered a priority intervention are the following:

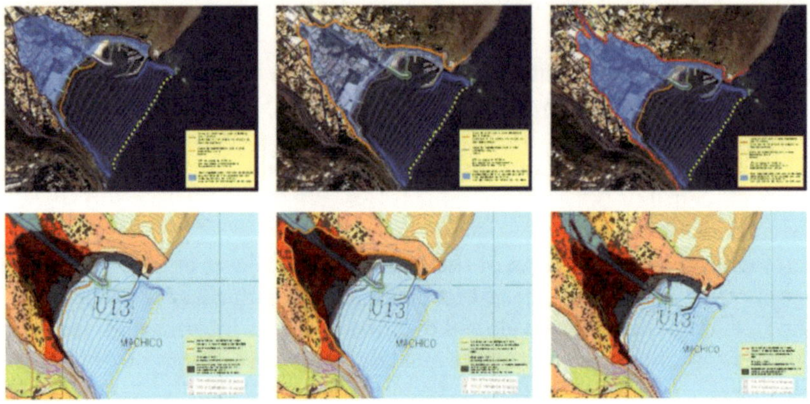

Figure 15.
Comparison of "MFL" maritime flood level – Hs (Real, Tr = 100 and Tr = 500) (source: Authors).

a. U1 – Machico old/historic area;

b. U2 – Machico equipment area;

c. U13 – Machico sea front.

After analyzing the Machico Master Plan, and in view of the increase in the value of Hsee, the uses of soil and their categorization were identified, with an increase of these in relation to the value of Hsee, as shown in **Figure 15**.

5. Closing section

The main focus of this study is the "Execution of an Artificial Beach and Respective Complementary Infrastructures (Madeira Island – Machico)."

From the bibliographic analysis, knowledge is acquired at the level of the execution methodologies related to the different alternatives to develop a complete artificial beach project.

So, considering the broad scope of the theme of this study and the slowness associated with it, it was possible to achieve the purposes previously stated in a satisfactory manner, complementing the knowledge acquired throughout academic life with scientific information and experimental analysis (use of different software).

In the development of this study, a comparative analysis between the "MFL" maritime flood level was carried out for the beach of Machico (South) – municipality of Machico (Portugal); as well as its iteration with the PDMM – essential instrument for spatial planning in the municipality of Machico, is processed, demonstrating that the creation of a project for a maritime work (artificial beach), is adequately connected with the territorial and urban planning, as well as the land uses described in the Machico Master Plan.

Two levels of risk of simple maritime flooding are distinguished:

1. The generated by the simultaneous performance of:

   a. Meteorological + astronomical + extreme weather tide – Whether the SEA storm (generated by cyclones or storms) or the SWELL (generated by rainstorms) causing maritime flooding, or by impact;

b. Astronomical + meteorological tide + wide waves – In this case, the greatest risk of flooding occurs when the astronomical and meteorological tides occur simultaneously.

2. The Flood is generated by extreme events at sea: tectonic, volcanic or impact tsunamis. These result from any combination of tides and are of such magnitude that it is not economically viable to design any infrastructure that minimizes damage.

3. The process for calculating the MFL is slow and laborious. However, it shows that an increase in the value of Hsee corresponds to the largest affected area on land, so greater area relative to different uses of soil will be affected.

Contextually, we believe this research would be essential to develop and improve some aspects if it is intended to continue to develop a similar study, namely:

a. The main aspect of being mentioned is the continuation and optimization of the monitoring carried out along the coast in RAM, in order to obtain a more accurate characterization of it, at different levels, undulation, geology, bathymetry, topography, etc.;

b. Optimize the spreadsheets developed throughout this study;

c. Modeling and interaction between wave heights and port structures, in order to minimize errors made over these years (recent past);

d. Analysis of the influence of the tide level in the flow in artificial water channels, direct relationship with the risk of floods downstream, and its interconnection with the analysis of the coastal dynamics (accounting of the effective solid transport).

Acknowledgements

This work was supported by national funds through the Fundação para a Ciência e a Tecnologia, I.P. (Portuguese Foundation for Science and Technology) by the project UIDB/05064/2020 (VALORIZA – Research Centre for Endogenous Resource Valorization).

Author details

Sérgio Lousada[1,2,3,4*], Luís Loures[2,5,6] and Rui Alexandre Castanho[2,4,7,8]

1 Faculty of Exact Sciences and Engineering (FCEE), Department of Civil Engineering and Geology (DECG), University of Madeira (UMa), Funchal, Portugal

2 VALORIZA – Research Centre for Endogenous Resource Valorization, Portalegre, Portugal

3 Institute of Research on Territorial Governance and Inter-Organizational Cooperation, Dąbrowa Górnicza, Poland

4 CITUR – Madeira – Centre for Tourism Research, Development and Innovation, Madeira, Portugal

5 Polytechnic Institute of Portalegre (IPP), Portalegre, Portugal

6 Research Centre for Tourism, Sustainability and Well-being (CinTurs), University of Algarve, Faro, Portugal

7 Faculty of Applied Sciences, WSB University, Dąbrowa Górnicza, Poland

8 College of Business and Economics, University of Johannesburg, Auckland Park, South Africa

*Address all correspondence to: slousada@staff.uma.pt

References

[1] K. Mangor, "General guidelines for good quality artificial beaches and Lagoons, and case stories," 2007.

[2] Lusa, "Madeira com praias para todos os gostos," 22 fevereiro 2020. [Online]. Available: https://www.dnoticias.pt/hemeroteca/268089-madeira-com-praias-para-todos-os-gostos-JODN268089.

[3] R. H. Turismo, "«Jardim inaugura praia artificial com areia de Marrocos»," 5 dezembro 2010. [Online]. Available: https://rhturismo.net/.

[4] Câmara Municipal de Machico, "Praias," 29 02 2020. [Online]. Available: https://www.visitmachico.com/pt/sentir/natureza-mar/praias.

[5] Peláez-Zapata, D. S., Osorio A. A. F., Álvarez-Silva, O., Osorio-Cano, J. D., Payán, L. (2012). Estimación de la Cota de Inundación como Herramienta para la Conservación de las Tortugas Marinas en el Pacífico Colombiano, XXV Congreso Latinoamericano de Hidráulica San José, Costa Rica, 9 al 12 de Septiembre de 2012. DOI:10.13140/2.1.5035.8402.

[6] Lousada, S. A. N., and Gonçalves, R. (2019). Metodologias de Determinação de Alturas de Onda. Dimensionamento de Obras Marítimas., ISBN 978-613-9-61105-8, Novas Edições Acadêmicas, Portugal, 253 pp.

[7] Gonçalves, R. S., Lousada, S. A., Castanho, R. A. (2017). Determinação de Alturas de Onda para Dimensionamento de Obras Marítimas. Article of Congress Book – I CIPSOT.

[8] Lousada, S. A. N. (2021). Underwater Work. Ed. IntechOpen. ISBN: 978-1-78985-229-5.

[9] Gonçalves, R. S., Camacho, R. F., Lousada, S., Castanho R. A. (2018). Modeling of maritime agitation for the design of maritime infrastructures: The case study of madeira archipelago (janeiro, 2018). Revista Brasileira de Planejamento e Desenvolvimento, Curitiba, v. 7, n. 1, p. 29-50. DOI:10.3895/rbpd.v7n1.7136.

Chapter 3

Frequent Change of Land-Use Pattern and Its Effect on Ecology and Ecosystem in the South-West Coastal Area of Bangladesh

Krishna Rani Barai

Abstract

The ecosystem of the South-West coastal area of Bangladesh is resourceful and unique. Regrettably, this distinctive area is vulnerable to numerous climatic factors, for instance, sea water intrusion, temperature variation, unpredictable behavior of rainfall, and frequent occurrence of cyclones and drought. Undesirably, the effects of those climatic factors have been enhancing on account of human behavior, one of which is the frequent change of land-use pattern in this area. The land-use pattern of the coastal Bangladesh has been changed continuously over the last 60 years in order to increase the agricultural production for improvement of livelihood. The unrealistic improvement projects (coastal embankment project, increased unscientific shrimp cultivation, and so on) brought the unforeseen disaster in the area. As a result, the salinity level of the area has been increased considerably, resulting in decreased production of different kind of crops and vegetables. Unstable economy and the emergence of calamities, leading to damage to ecology and ecosystem, made livelihood insecure in the area.

Keywords: land-use change, salinity, shrimp farming, coastal area, ecology, ecosystem

1. Introduction

Land degradation is an urgent environmental issue in many countries. Frequent changes in land-use pattern and soil salinity have contributed to land degradation in the coastal area of Bangladesh. This chapter is intended to unfold the scenario of land-use pattern changed over a long period in the South-West coastal area of Bangladesh and how it influenced the soil salinity by practicing shrimp farming. In addition, this chapter describes the ecological and socioeconomic conditions of the area in response to those changes. Furthermore, it includes some policy recommendations and management approaches to maintain productive agriculture in the area.

2. Land use and land cover (LULC)

Land-use refers to how people are using the land, for example, settlement, agriculture, mining, etc., whereas land cover refers to the surface cover on the ground, such as forest, vegetation, open water, bare soil, etc. Land-use includes the management and modification of natural environment or wilderness into manufactured environment, for example, settlements and seminatural habitats like productive fields, pastures, and managed woods.

3. LULC change (LULCC)

Land-use and land cover change (LULCC) is the conversion of different types of land-use and is the result of complex interactions between humans and the physical environment. LULCC might be a major driver of global alteration and has a considerable impact on ecosystem processes, biological cycles, and biodiversity.

4. Importance of coastal Bangladesh

4.1 The coast of Bangladesh

Bangladesh coast consists of 19 districts which cover 32% of the total area of the country [1]. The population in the coastal area is expected to increase from 36.8 million in 2001 to 60.8 million by 2050 [2]. This coastal belt has miscellaneous natural resources counting the largest mangrove forest in the world (Sundarbans), salt, coastal fisheries, and other minerals. Additionally, this part of the country has a high exploration prospective for both onshore and offshore natural gas [3]. The resourceful ecosystems of this coastal area maintain the livelihoods of the local communities. Moreover, this region has the potential for tourism, ports, and other developments [4]. But this region is highly exposed to various climatic factors, such as temperature fluctuation, erratic behavior of rainfall, increased frequency of cyclones, drought, and saltwater intrusion [5].

4.2 Agriculture in the coastal area of Bangladesh

Agriculture is the major sector in the economy of Bangladesh, and over 30% of the net cultivable land is in the coastal area [6]. In the past, people were interested to cultivate not only local varieties that are saline tolerant but also other varieties with greater plant height and comparatively low production cost and are tasty and above all easily manageable. The local rice varieties are categorized into two types, one for planting in the freshwater shrimp farms and another for other agricultural lands. *Jotabalam* and *ghunshi* varieties are selected to be cultivated in the shrimp farms. On the other hand, *ashfall* and *benapol* varieties are designed for planting in other agricultural lands.

5. Salinity issue

Soil salinity is one of the most common causes of soil degradation, which is getting worse day by day [7]. Usually soil salinity refers to surface or near-surface salt accumulation [8]. It is defined as the concentration of dissolved salts, mainly chlorides and sulfates, in soil or water [9]. It is practically expressed as practical salinity unit (PSU) which depends on the water temperature and conductivity [10]. It is also expressed as ppt (parts per thousand). The characterization of soil salinity is normally done by spectrometer through measuring the electric conductivity (EC) in a saturated soil paste or in aqueous extracts with different soil/water ratios [11].

5.1 Soil salinity situation in the coast of Bangladesh

Out of 2.85 million hectares of the coastal and offshore lands, about 1.05 million hectares of arable lands are affected by varying degrees of salinity [6]. From another study, it is found that the salinity-affected area has increased from 8330 km^2 in 1973 to 10,560 km^2 in 2009, of which about 4530 km^2 is affected by a higher level of salinity (more than 8 dS/m), indicating 43% of the total salt-affected areas are facing challenges for agricultural practices even with salt-tolerant rice varieties [12]. Huge economic loss, damage to rice crops, and difficulties in drinking water supply systems in many villages of the coastal districts are the common problems caused by salinity [5].

Salinity is one of the most significant problems in the South-West coastal belt which is denoted as a food-deficit area of Bangladesh. Both the diversity of food and the net food production have declined significantly over the last decades [13]. It is reported that the worst salinity conditions are found in Khulna, Bagerhat, Satkhira, and Patuakhali districts [12]. Around 70% of the land in Barisal and Khulna divisions was affected by different degrees of salinity [14].

5.2 Major causes of salinity in the coastal area

Generally salinity can develop naturally (primary salinity), but human interference accelerated the movement of salts into rivers and onto land (secondary salinity). The spatial inconsistency of soil salinity over the landscape is highly sensitive and controlled by a variety of factors: soil factors (parent material, permeability, depth of water table, groundwater quality, and topography), management factors (irrigation and drainage), and climatic factors (rainfall, temperature, and humidity) [15]. Hence it is difficult to assess human interference on land salinity separately as it is highly integrated with environmental complexity [16].

Secondary salinity is mostly related to inequitable exploitation of natural resources, global climate change, and land-use changes, such as overgrazing and excessive utilization of land and water resources [17]. Land-use changes disturb hydrologic conditions which in turn alter the land salinity pattern of the area [18]. Thus poor land and water management system in irrigated farmlands leads to secondary salinity worldwide [19].

Bangladesh is one of the most vulnerable countries in salinity issue. The major causes of saline intrusion in water and soil include cyclone and storm surge, sea level rise, and shrimp farming practices [5]. Thus, shrimp farming is one of the major causes of secondary salinity in Bangladesh.

6. How shrimp farming influences the economy and LULC in the coastal Bangladesh

Shrimp industry is mostly concentrated in tropical developing countries making shrimps as a major export item for the Western countries [20]. This industry has experienced amazing growth over the last decades [13]. In the Asia-Pacific region, it is one of the highest growing economic activities [21].

Shrimp farming plays an important tool for poverty reduction and livelihood improvement for some households, but it has an adverse impact on coastal environment [22, 23], particularly in Southeast Asia. Two major environmental impacts of shrimp culture are the consumption of resources and the subsequent release of wastes into the environment [24, 25]. Other impacts include reduced water flow, soil salinization, diminution and salinization of ground- and surface water, depletion of wild fish and shrimp populations, extinction of wild vegetation, and biological pollution of native shrimp stocks [20, 22, 23]. The topsoil (0–20 cm) of the shrimp ponds are mostly affected by salinity, and it is decreasing further inland [26]. In Asia, shrimp farming has been far away from success as the industry claims, especially when environmental costs are internalized [27].

Shrimp is denoted as the “golden price” in Bangladesh, because of its valuation in the international market. Shrimp production got the position as the second largest export business in Bangladesh by earning US$456 million in 2006 [28]. The fisheries’ part, together with shrimp, contributes about 6% of the GDP and 5% of the national export earnings, with shrimp alone contributing about 93% of the sectoral export earnings [29]. This enlighten why the coastal areas were declared as a “free zone” for shrimp cultivation in 1994 by the Bangladesh government [30]. Subsequently, shrimp cultivation extended noticeably. In favor of example, shrimp was cultivated on 39% of the land of the Shyamnagar *Upazila* (subdistrict) of Satkhira District in 2002. It had enlarged to 57 and 62% by 2007 and 2011, respectively [13]. It is also reported in a recent analysis that 83% of the paddy field of 1988 was converted to different other land-use type in 2017, of which 23% is converted to water bodies [31].

7. LULC situation in the South-West coastal area of Bangladesh

7.1 Land and land-use pattern in Bangladesh

Land is the basic natural resource that provides not only the habitat and nourishment for living organisms but also a major focus of economic and livelihood activities. With the increasing population, land is being converted from agricultural purposes to other uses (such as housing, roads, and urban development), and this trend is expected to continue. Around 220 ha of arable land are converted to industrial establishment, house, road construction, etc. every day [32]. Between 1973 and 2000, at least 86,000 ha of land were lost into river/estuarine erosion though this is compensated by land generated through accretion [33].

The Bangladesh Bureau of Statistics regularly publishes land-use pattern where importance is mainly focused on agriculture. Here, land-use system is usually determined by physiography, climate, and land height in relation to water level [34]. All these together make a highly complex environment characterized by five main land types: net cropped area, current fallow, current waste, forest, and area not available for cultivation. Besides this, the Soil Resources Development Institute (SRDI), Bangladesh, produces agricultural land-use maps for the country identifying many different types of agricultural land-use. An indicative land zoning consisting of the following eight zones was recommended by PDO-ICZMP [35]:

1. Agricultural zone
2. Shrimp (brackish-water) zone
3. Shrimp (sweet-water) zone
4. Salt-shrimp zone
5. Forest zone
6. Mangrove (including Sundarbans) zone
7. Urban and commercial zone (industrial, port, export-processing zones, and ship breaking yards)
8. Tourism zone

7.2 How the LULC changes occurred in the coastal area of Bangladesh

Over the last 60 years, the land-use pattern of the coastal Bangladesh has changed repeatedly. In agricultural sector, rice was the main crop in the area, in the 1950s, [6]. However, rice production decreased regularly, because of tidal flooding and rising salinity [36]. With the help of the World Bank and other funding bodies, the Coastal Embankment Project was established between 1960 and 1980 to increase rice production in the coastal areas [37]. Regrettably, this project formed harmful impacts on the ecosystems of the area. Even though rice production increased immediately after the construction of embankments, the production had dropped another time by the 1990s. The major reason behind this decreased production was the congestion throughout the drainage system in the land area behind the embankments due to poor maintenance and inadequate management [38]. Consequently, the rice cultivators changed their livelihood to shrimp (black tiger) cultivation. This change of land-use again smashed the ecosystems of the area, as paddy fields and several parts of the mangrove forest were transformed to shrimp farm [39]. Therefore, the salinity of the area increased noticeably because of the enlarged shrimp cultivation. This change had an adverse effect on crop production as well as local vegetation, fish, and plant diversity over the last decade in the area [13].

Moreover, 30-year satellite data demonstrates that paddy field was the foremost category (37%) in the 1980s, which is continuously decreased to 9% in 2017. On the contrary, water bodies were the least dominant category in the 1980s (17%), but they constantly enlarged to be a dominant category in 2017 (34%) (**Figure 1**). There is another interesting finding from that research, where each of the LULC categories of 2017 was again split to perceive the land-use type in the 1980s. Thirty four percent of the water bodies in 2017 were water bodies in 1988, 26% water bodies of 2017 were directly transformed from paddy fields, and 40% were indirectly transformed through homestead gardens and settlements/bare land [31]. Therefore,

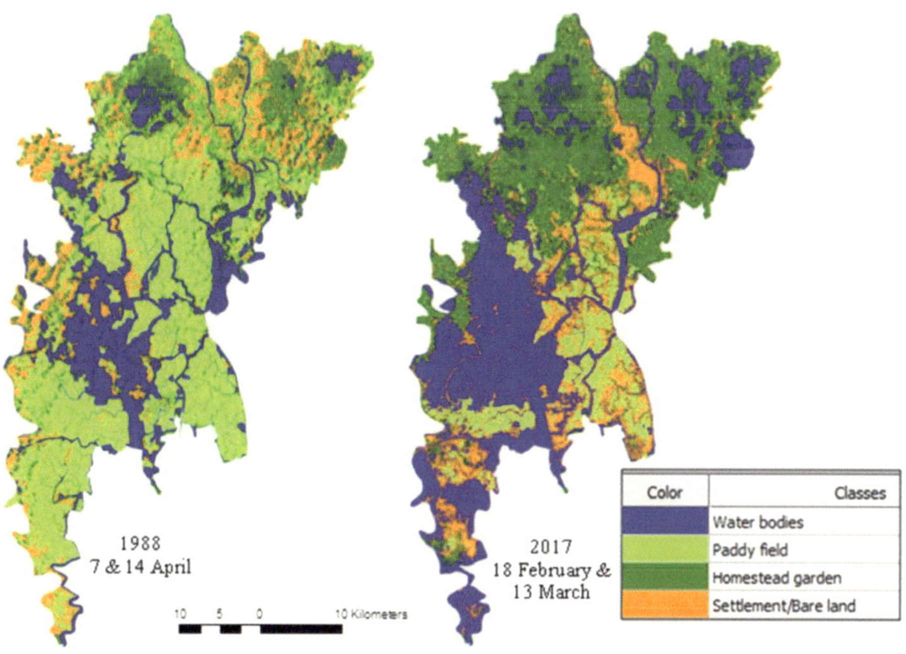

Figure 1.
LULC classification map of Khulna District in 1988 and 2017 (source: [31]).

it is evident that the LULC is changed frequently in the coastal area of Bangladesh, which has a direct effect on the ecology and ecosystem.

7.3 Causes and effects of the LULC changes

There are many reasons of the frequent changes of LULC in the coastal Bangladesh. Two of those are:

1. **Increasing salinity level:** the reason behind increasing salinity level are:

    - The presence of the Bay of Bengal.
    - Lack of proper management of agricultural land and irrigation system
    - Lack of proper management and dredging of river reduce the current of the river, which in turn increases seawater intrusion both in surface water and groundwater gradually. As a result, salinity covers more area progressively.
    - Moreover, soil porosity is reducing in this area. Therefore, groundwater infiltration is lowering, which increases waterlogging in the area. As a result, both the soil and water salinity is increasing continuously. Three major reasons reducing soil porosity are:

        i. The use of huge amount of polythene

        ii. Shrimp cultivation: sometime shrimp cultivators are using salt in the shrimp cultivation pond to avoid some virus and other diseases. This salt creates an invisible layer in the soil.

        iii. Some projects like tree plantation (eucalyptus tree, etc.). Leaves of those trees are not easily decomposable, which create a layer in the soil. Consequently, the porosity is reducing.

2. **Decreasing agricultural production:** agricultural production is decreasing because of two major reasons:

    - Due to poor management of the embankment and sluice gate that was built in the 1960s to the 1980s.
    - Some paddy fields were converted to shrimp cultivation ponds [31]. As a result, the salinity level of the surrounding paddy field of the shrimp cultivation pond increased, which lead to reduced agricultural production.

Frequent changes of LULC have a long-term effect on the paddy field and other vegetation, reducing productivity, biodiversity, and wild varieties.

7.4 How the ecological and socioeconomic condition of the area responded to those changes

LULC in the coastal area of Bangladesh had been changed frequently since the 1960s. The major agricultural conversions were rice to shrimp and shrimp to rice cultivation. Even though both rice and shrimp cultivation create vital contributions to the GDP of Bangladesh, unregulated shrimp cultivation has had adverse ecological impacts in the area. Shrimp cultivated area like Tildanga *Union* of Dacope *Upazila* in Khulna District lost the ecological balance. Agricultural production along with trees and other vegetation had noticeably declined in the area. Furthermore, both the domestic and wild animals were reduced considerably [31]. This ecological imbalance has a direct and indirect effect on the mass economy and community health of the area. The circumstances are getting worse day by day.

On the contrary, Kamarkhola is a *union* with a similar situation as Tildanga. Therefore, most of the shrimp farmers changed their major livelihood to rice cultivation as both the ecology and economy of the area were destroyed visibly. As a result of shifting occupation, the ecology and the economy of the area got balanced within 10 years. Additionally, the sources of income are more diversified in Kamarkhola, at present, which may decrease the risk of livelihood [31].

8. Policy recommendation and management approaches

There are some recommendations:

1. Designing a proper and scientific coastal zoning system. This will be ecologically and economically feasible to get optimum production to meet the country needs.

2. Formulating a typology for each zone explaining its characteristics in detail.

3. Planning a cropping pattern of the area by analyzing the seasonal salinity situation for the betterment of the production, because soil salinity varies considerably in each season.

4. Searching an appropriate land-use system in the coastal area through the technical support of RS-GIS.

9. Conclusions

Naturally, LULC will change over the time for the seek of development. If LULC is altered totally and frequently, it will damage the ecology and ecosystem of the area. Consequently, the nature will be imbalanced, causing disaster for the area. Therefore, we should use the technology to design the proper land-use pattern considering the natural value of that area rather than going against the nature. Accordingly, both the nature and human being will get benefited from this technology-based planning, and thus, the development will be sustainable.

Author details

Krishna Rani Barai
LoCOS - Loving Care for the Oppressed Society, Batiaghata, Khulna, Bangladesh

*Address all correspondence to: krish_051002@yahoo.com

References

[1] Huq S, Rabbani G. Adaptation technologies in agriculture; the economics of rice farming technology in climate–vulnerable areas of Bangladesh. In: Christiansen L, Olhoff A, Traerup S, editors. Technologies for Adaptation: Perspectives and Practical Experiences. UNEP Riso Centre, Roskilde. Roskilde, Denmark: UNEP Riso Centre on Energy, Climate and Sustainable Development; 2011. Available from: http://www.unep.org/pdf/TechnologiesAdaptation_PerspectivesExperiences.pdf

[2] PDO-ICZMP (Program Development Office for Integrated Coastal Zone Management Plan). Living in the Coast: Urbanization. Dhaka, Bangladesh: PDO-ICZMP, Water Resources Planning Organization; 2005. p. 36

[3] Michel D, Pandya A. Coastal Zones and Climate Change. Stimson's Regional Voices. Washington, DC: The Henry L. Stimson Center; 2010. Available from: www.stimson.org/programs/regional-voices [Accessed: 12 October 2016]

[4] Uddin MS, Shah MAR, Khanom S, Nesha MK. Climate change impacts on the Sundarbans mangrove ecosystem services and dependent livelihoods in Bangladesh. Asian Journal of Conservation Biology. 2013;**2**(2):152-156

[5] Rabbani G, Rahman A, Khandaker M, Shoef IJ. Loss and damage from salinity intrusion in Sathkira District, coastal Bangladesh. In: Loss and Damage in Vulnerable Countries Initiative, Case Study Report. Bonn: United Nations University Institute for Environment and Human Security; 2013

[6] Asib A. Some of the major environmental problems relating to land use changes in the coastal areas of Bangladesh: A review. Journal of Geography and Regional Planning. 2011;**4**(1):1-8

[7] Jordán M, Navarro-Pedreño J, García-Sánchez E, Mateu J, Juan P. Spatial dynamics of soil salinity under arid and semi-arid conditions: Geological and environmental implications. Environmental Geology. 2004;**45**:448-456

[8] Tanji K. Salinity in the soil environment. In: Lauchli A, Luttge U, editors. Salinity: Environment-Plants-Molecules. Dordrecht, The Netherlands: Kluwer Academic Publisher; 2004. pp. 21-51

[9] Smith DI, Finlayson B. Water in Australia: Its role in environmental degradation. In: Heathcote RL, Mabutt JA, editors. Land, Water and People: Geographical Essays in Australian Resource Management. North Sydney: Allen and Unwin; 1988

[10] Fofonoff NP. Physical properties of seawater: A new salinity scale and equation of state for seawater. Journal of Geophysical Research. 1985;**90**(C2):3332-3342

[11] Ben-Dor E, Goldshleger N, Eshel MV. Combined active and passive remote sensing methods for assessing soil salinity: A case study from Jezre'el valley, northern Israel. In: Metternicht G, Zinck JA, editors. Remote Sensing of Soil Salinization: Impact on Land Management. Boca Raton, FL, USA: CRC Press; 2008. pp. 236-253

[12] SRDI (Soil Resources Development Institute). Saline Soils of Bangladesh. Dhaka, Bangladesh: SRDI, Ministry of Agriculture; 2010

[13] Islam MS, Rahman MA, Sultana N, Nath B, Paul A. Using geospatial techniques to assess the salinity impact on agricultural land use: A study on Shyamnagar Upazila, Satkhira. Journal of

Agriculture and Environment for International Development—JAEID. 2012;**106**(2):157-169

[14] Rahman M, Ahsan M. Salinity constraints and agricultural productivity in coastal saline area of Bangladesh. In: Soil Resources in Bangladesh: Assessment and Utilization. Proceedings of the Annual Workshop on Soil Resources, 14-15 February 2001. Dhaka, Bangladesh: Soil Resources Development Institute; 2001. pp. 1-14

[15] Douaik A, Meirvenne M, Toth T. Stochastic approaches for space–time modeling and interpolation of soil salinity. In: Metternicht G, Zinck JA, editors. Remote Sensing of Soil Salinization: Impact on Land Management. Boca Raton, FL, USA: CRC Press; 2008. pp. 273-289

[16] Zhang TT, Zhao B. Impact of anthropogenic land-uses on salinization in the Yellow River Delta, China: Using a new RS-GIS statistical model. In: International Archives of the Photogrammetry, Remote Sensing and Spatial Information Science, Volume XXXVIII, Part 8. Kyoto Japan: International Archives of the Photogrammetry, Remote Sensing and Spatial Information Science; 2010

[17] Li XY, Wang ZM, Song KS, Zhang B, Liu DW, Guo ZX. Assessment for salinized wasteland expansion and land use change using GIS and remote sensing in the west part of northeast China. Environmental Monitoring and Assessment. 2007;**131**:421-437

[18] George RJ, McFarlane DJ, Nulsen RA. Salinity threatens the viability of agriculture and ecosystems in western Australia. Hydrogeology. 1997;**5**:6-21

[19] Metternicht GI, Zinck JA. Remote sensing of soil salinity: Potentials and constraints. Remote Sensing of Environment. 2003;**85**:1-20

[20] Thornton C, Shanahan M, Williams J. From Wetlands to Wastelands: Impacts of shrimp farming. Environmental Justice Foundation. SWS Bulletin; 2003. pp. 48-53

[21] Wolanski E, Spagnol S, Thomas S, Moore K, Alongi DM, Trott L, et al. Modelling and visualizing the fate of shrimp pond effluent in a mangrove-fringed tidal creek. Estuarine, Coastal and Shelf Science. NA: Elsevier Science; 2000;**50**:85-97

[22] Nhuong TV, Luu LT, Tu TQ, Tam PM, Nguyet TTA. Vietnam shrimp farming review. In: Individual Partner Report for the Project: Policy Research for Sustainable Shrimp Farming in Asia. European Commission INCO-DEV Project PORESSFA No.IC4-2001-10042, CEMARE University of Portsmouth UK and RIA1: Bac Ninh, Vietnam. 2002

[23] EJF (Environmental Justice Foundation). Risky Business: Vietnamese Shrimp Aquaculture—Impacts and Improvements. London: Environmental Justice Foundation; 2003

[24] Beveridge HCM, Phillips MJ, Macintosh DJ. Aquaculture and the environment: The supply of and demand for environmental goods and services by Asian aquaculture and the implications for sustainability. Aquaculture Research. 1997;**28**:797-808

[25] Kautsky N, Beveridge M, Folke C, Primavera JH, Ronnback P, Troell M. Aquaculture and biodiversity. In: Levin S, editor. Encyclopedia of Biodiversity. Vol. 1. London: Academic Publishers; 2000. pp. 185-198

[26] Tho N, Vromant N, Hung NT, Hens L. Soil salinity and sodicity in a shrimp farming coastal area of the Mekong Delta, Vietnam. Environmental Geology. 2008;**54**:1739-1746

[27] Ronnback P, Bryceson I, Kautsky N. Coastal aquaculture development in

eastern Africa and the western Indian Ocean: Prospects and problems for food security and local economies. Ambio. 2002;**31**:537-542

[28] Bangladesh Frozen Foods Exporters Association (BFFEA). Shrimp and Fish News, April–June, 2008. Dhaka, Bangladesh: BFFEA; 2008. pp. 2-41

[29] Department of Fisheries, Bangladesh. Annual Report (2005-2006), Dhaka Bangladesh; 2007. pp. 27-51

[30] Rahman A, Ali L, Mallick DL. An overview and perspectives on the Bangladesh shrimp industry. In: Rahman A et al., editors. Shrimp Farming and Industry: Sustainability, Trade and Livelihoods. Dhaka: The University Press; 2006

[31] Barai KR, Harashina K, Satta N, Annaka T. Comparative analysis of land use pattern and socioeconomic status between shrimp- and rice-production areas in southwestern coastal Bangladesh: A land-use/cover change analysis over 30 years. Journal of Coastal Conservation. 2019;**17**(2). DOI: 10.1007/s11852-019-00682-2

[32] Islam MS, Razzaque MA, Rahman MM, Karim NH. Present and future of agricultural research in Bangladesh (in Bengali). In: Agriculture in the 21st Century: Challenges and Possibilities. Ministry of Agriculture, Dhaka, Bangladesh; 2004. pp. 20-27

[33] MES (Meghna Estuary Study). Hydro-Morphological Dynamics of the Meghna Estuary. Dhaka, Bangladesh: MES Project, Bangladesh Water Development Board; 2001

[34] Brammer H. Land Use and Land Use Planning in Bangladesh. Dhaka, Bangladesh: The University Press Limited; 2002. p. 554

[35] PDO-ICZMP (Program Development Office for Integrated Coastal Zone Management Plan). Coastal Land Uses and Indicative Land Zones. Dhaka, Bangladesh: PDO-ICZMP, Water Resources Planning Organization; 2005. p. 64

[36] Nishat A. Review of present activities and state of art of the coastal areas of Bangladesh. In: Coastal Area Resource Development and Management. Part II, Coastal Area Resource Development and Management Association (CARDMA), Dhaka Bangladesh. 1988. pp. 23-35

[37] Program Development Office for Integrated Coastal Zone Management Plan. Where Land Meets the Sea: A Profile of the Coastal Zone of Bangladesh. Dhaka, Bangladesh: The University Press Ltd; 2004

[38] Rahman A. Beel Dakatia: Environmental Consequences of a Development Disaster. The Dhaka, Bangladesh: University Press Ltd; 1995

[39] Haque AKE. Sanitary and Phyto-Sanitary Barriers to Trade and its Impact on the Environment: The Case of Shrimp Farming in Bangladesh. Dhaka Bangladesh: IUCN Bangladesh Country Office; 2004. p. 63

Chapter 4

Using the Buffer Zone Method to Measure the Accessibility of the Green Areas in Tokat, Turkey

Mustafa Ergen

Abstract

This study measured the distance from its sample location to green areas in Tokat, Turkey, using the buffer zone method. This study evaluated the decisions of the Purposeful Tokat Conservation Framework Development Plan and the decisions outside of the Historical Urban Protected Areas Development Plan together as whole. Creating green areas that are at an accessible distance in urban planning projects is important. Accordingly, the accessibility of green areas was measured using the buffer zone method. This study also proposes a method in regard to accessibility to green areas in Turkey. The analysis performed in Tokat found that the green areas were not planned considering accessibility criteria. The results of this study indicate that settlements with houses and green areas should be planned proportionally and at a more accessible distance to each other.

Keywords: Tokat, urban planning, buffer zone method, accessibility, green areas

1. Introduction

Increased population is an important factor in the formation of urban areas within the human settlements for a very long time [1]. During the development of urban areas, settlements maintained some of their rural qualities until the industrial revolution. After that, the formation of urban areas transformed rural areas into industrial zones [1]. Urban areas had many problems in this period of development, which were joined by more problems with higher priority. Planning green areas to make them accessible has been an important issue since the beginning of urbanization, and suggestions and practices have been introduced along with legal regulations to ensure that green areas are planned, healthy, and sustainable.

As access to urban green areas has become a priority for the modern urban areas, increasing quality of life and creating open areas with proper airflow have also become important. Morar et al. indicated the contributions of green areas to urban quality of life [2]. Comber et al. helped measure the accessibility of green areas for different ethnic groups [3].

The term, green areas, covers many fields and meanings. Green areas may indicate recreational areas, playgrounds, or sports fields. The term refers to locations that increase quality of life and help people breathe fresh air and have space for leisure time and fun.

Urban planning refers to the approaches needed to organize and create healthy urban areas. Ergen explained that urban planning means making arrangements that help people to establish orderly relationships with their environments [4]. Accordingly, efforts have been made to conduct sustainable and healthy urbanization and green area arrangements through planned approaches.

Remarkable new approaches and arrangements have been performed within urban planning in accordance with the development of urban areas. Efforts have also been made to provide the best recommendations for resolving urban problems. With developments in dynamic urban structures, many predictable and unpredictable social issues have arisen. To create a more livable environment within the urban areas, 7 m^2 of green space per person was allocated for green areas in Turkey, and this was later updated to 10 m^2 per person. The new 2017 regulation on spatial planning raised this figure to 15 m^2 for actively used green areas in districts and provinces.

These regulations provide little guidance for arranging green areas. The first regulation on spatial planning (2014) indicated that green areas should be designed within 500 m from residences to ensure spatial quality and meaningful planning. However, the regulation contains no recommendations or methods on how to measure the 500 m in question. This study used the buffer method, which has often been used in the international literature, to measure accessibility in Turkey.

This method was used in the English cities, e.g., Norfolk, to measure distances to green areas [5]. In 2017, Kolitos and Papadopoulou measured the accessibility of green areas using the buffer method [6]. The accessibility of green areas is an important topic for everyone. For instance, green areas that are not accessible to the young, elderly, or disabled people do not adequately fulfill their function and indicate poor design and planning.

Enabling people to implement the buffer method, ArcGIS is a significant instrument. Many scientific studies are based on a scientific ground using ArcGIS. Its useful practices include land suitability maps and planning approaches and analyses [7–11].

The importance of understanding green areas' accessibility is clear. The active use of green areas requires them to be accessible, and since unused green areas are not meaningful, the active use of green areas, including open areas, recreational areas, and parks, is important.

Tokat is a city which has historical infrastructure with a conservation development plan. It is sure that the arrangements of green areas are difficult in conservation areas, because the land is valuable and the development is under control for keeping alive historical structures in the area. In this perspective, it is needed to construct green areas that should serve to the conservation areas within the entire city. It is sure that the buffer zone method appears as one of the most suitable techniques for the conservation areas in the cities.

The buffer zone method was used in this study to measure the accessibility of green areas in Tokat, considering the 500-m requirement in Turkey. This method is an important procedure for assessing the accessibility of green areas. This study's results show the problems and advantages of green areas that were designed before the adoption of the 500-m rule, and what this rule will do for urban planning was demonstrated.

2. Methodology

2.1 Study area and settings

The province of Tokat has border with the provinces of Yozgat, Amasya, Samsun, Ordu, and Sivas. It is located in the Central Black Sea Region at northern latitude 39°

52′-40° 55′ and eastern longitude 35° 27′-37° 39′ [12]. According to data from the Turkish Statistical Institute (TSI), its address-based population is 166,136 [13] (**Figure 1**).

The development plans for Tokat were made using NetCAD and later transformed into ArcGIS data. This study uses its 1/5000 development plan and shows how its urban development emerged over the suggested fields. The development plans for Tokat were also put into the WGS 1984 Universal Transverse Mercator (UTM) Zone 35 North in ArcGIS to use the maps more effectively. Buffer zones were drawn using ArcGIS, and the distance of city's buildings to green areas was measured considering the 500 m rule.

The arrangements of green areas in the framework of Tokat conservation plan should bring to the fore structure of historical buildings. If this arrangement cannot be done, it is necessary to enrich green areas nearby to the conservation areas. It is important to reveal green areas with buffer zone in a way to provide integrity with conservation areas (**Figure 2**).

The data were classified. Housing zones and recommended housing zones were assessed together. The forest zones that were believed to host active green areas were grouped as parks, playgrounds, and woodlands (the recommended woodlands are also in these areas), and their accessibility was measured.

Although the term, buffer zone, has been used for a long time (occasionally for multiple use or transitional zones), its recent use as a concept date to the early 1970s [14]. Having been used to refer to the permitted vicinity of natural reserves, the term is now commonly used in many fields.

Buffer zones determine the limits of approaching and protecting natural reserves. The buffer zone method can be used to measure the usability and accessibility of urban open green areas. The importance of locations in buffer zones becomes clear based on usage purposes. The impact of use by people on natural reserves is measured to protect nature. The buffer zone method is an analytical method for understanding how green areas in urban buffer zones should have balanced usage and how deficiencies should be met (**Figure 3**).

The map of Tokat indicates that there are no housing zones on the west side of the city and that urban development has mainly occurred on the north, south, and

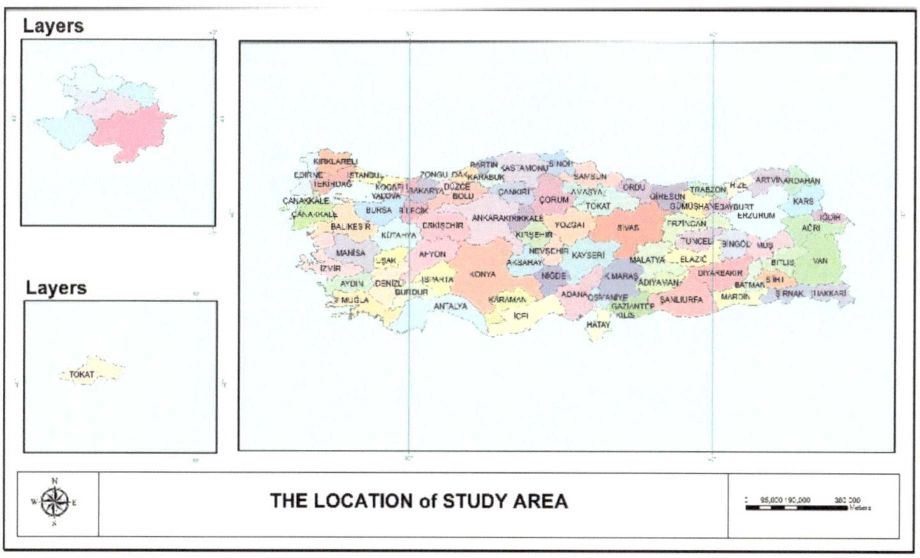

Figure 1.
The location of study area.

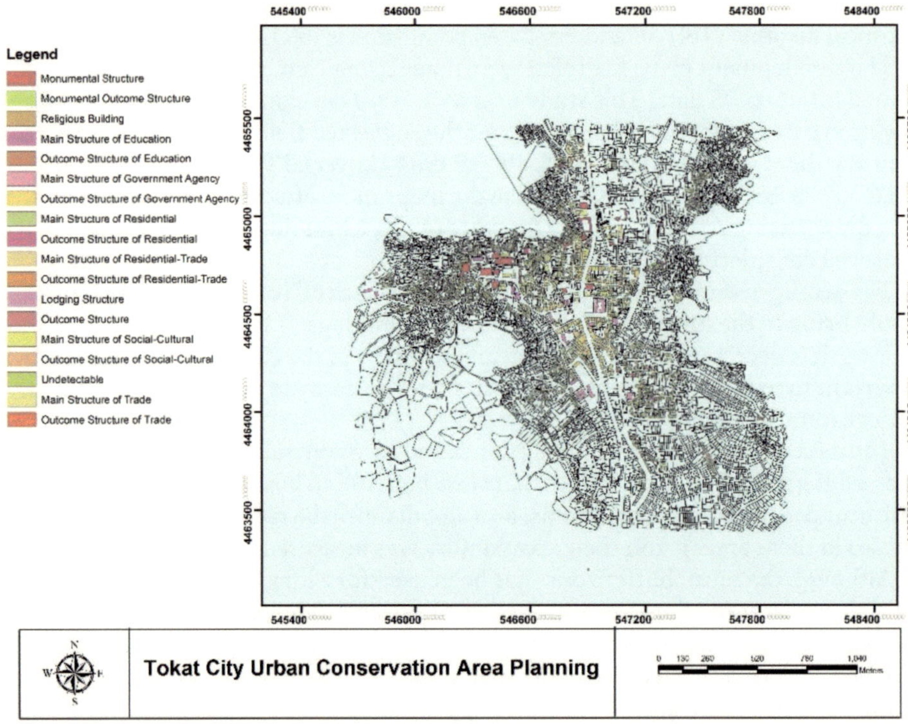

Figure 2.
Source: The map was obtained from Tokat municipality with ArcGIS program. The data is arranged by author for analyzing Tokat city.

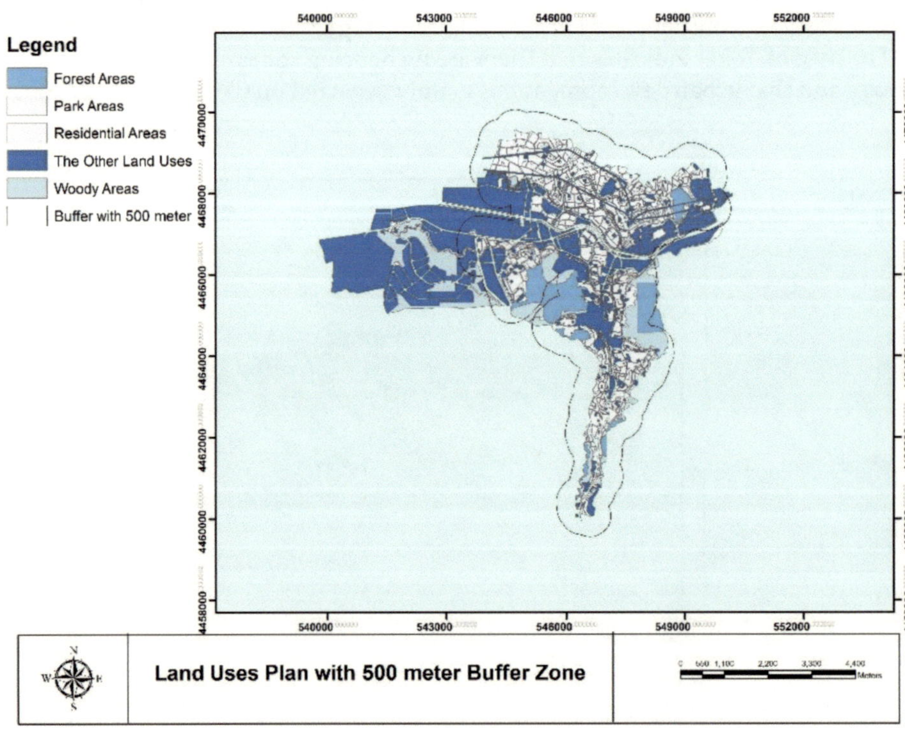

Figure 3.
Source: The map was obtained from Tokat municipality with NetCAD program. The map was converted to ArcGIS and prepared by the authors for description of analyzing areas in Tokat.

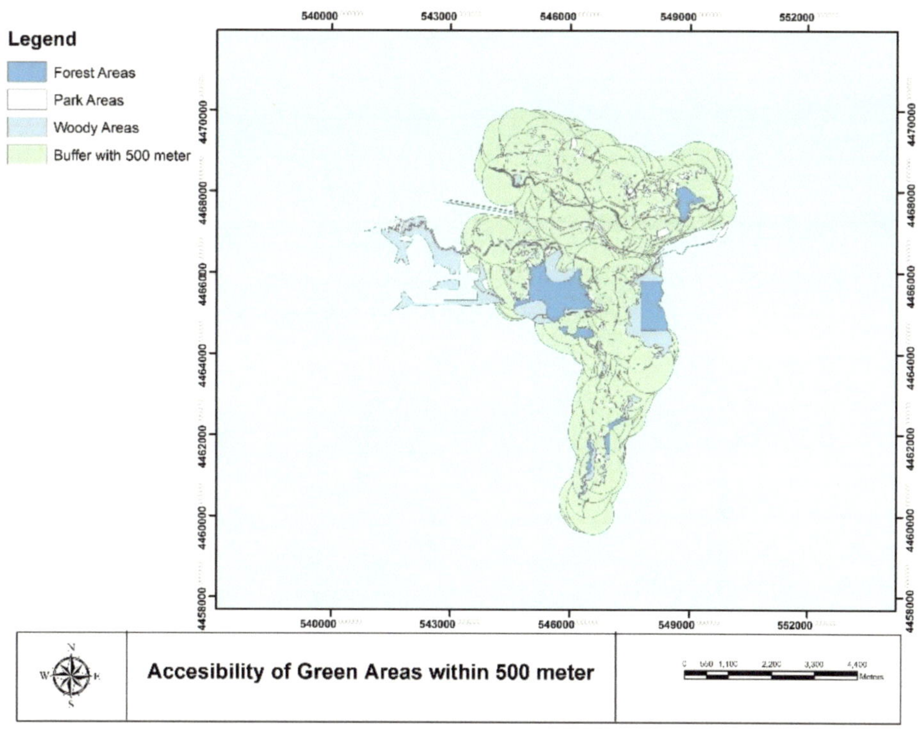

Figure 4.
Source: The map was obtained from Tokat municipality with NetCAD program. The map was converted to ArcGIS and prepared by the authors for description of analyzing areas in Tokat.

east sides. A green area planning activity on the west side where other terrestrial uses are common draws attention. The fact that these green areas do not serve any housing zones requires a reconsideration of the Tokat's planning decisions (**Figure 4**).

The map shows that the green areas have been arranged without considering accessibility. The green areas outside the distance contribute to the number of green areas per capita, but they do not mean anything in terms of accessibility. The fact that no recommended housing zones are in this zone indicates that its green areas were not formed with a planning-based approach for accessibility. In addition, green areas were found to have been designed and planned without forming a unity. The map also shows that the green areas were designed inconsistently without considering planning principles.

3. Result

Of all the entire green areas, 53% were in the buffer zone, including 44% of park areas, 61% of forests, and 54% of woodlands. Although the accessibility of the green areas seemed to be ensured in this zone, 56% of the parks were not designed for accessible distances in a planned manner. The green areas were proposed to the region of the centers of the city in the old development plan. This planning decision exposed insufficient green areas and increased to predict 15,200 m^2 green areas in the solution approach [15] (**Table 1**).

The results indicate that the green areas in the city were distributed without planning criteria. The fact that most of the green areas are not within accessible distance indicates that Tokat's planning decisions should be reconsidered. The presence

Unit of green spaces	Percentage of areas in 500 m
Park areas	44%
Forest areas	61%
Woody areas	54%

Table 1.
The percentage of green areas in buffer zone.

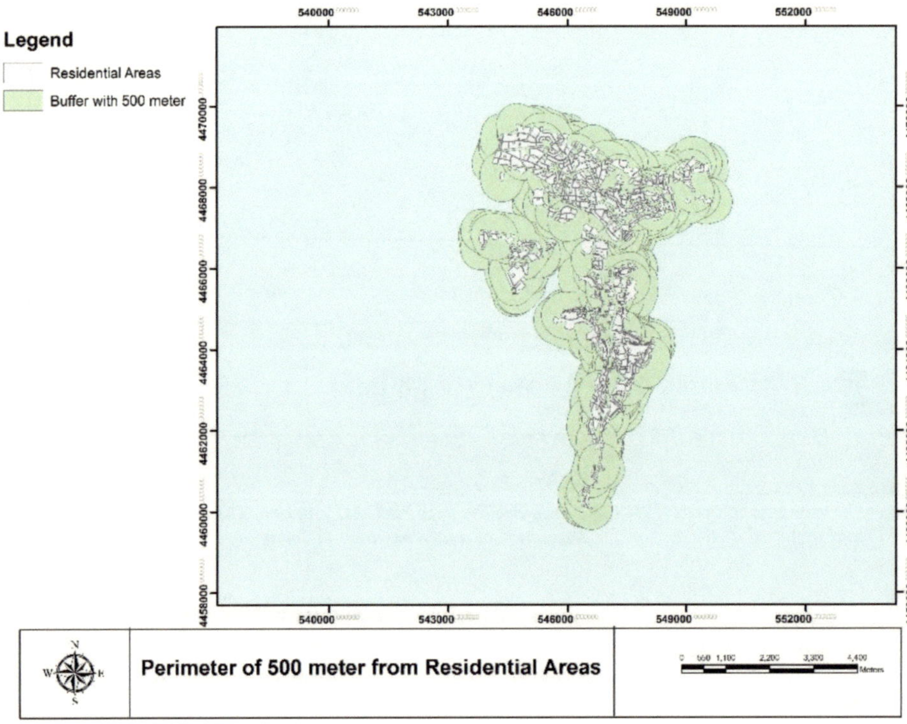

Figure 5.
Source: The map was obtained from Tokat municipality with NetCAD program. The map was converted to ArcGIS and prepared by the authors for description of analyzing areas in Tokat.

of woodlands and forests also affected the city's planning decisions. The important point here is how much of the forests and woodlands are in active use (**Figure 5**).

Considering the distance from the housing zones to green areas, the 500-m rule was generally implemented. However, the green areas in the city do not have any unity, and not all of them are within 500 m from housing, which shows a case of unpracticality for young, elderly, and disabled people and raises the problem of their accessibility. Although there are numerous green areas around the housing zones, most of them are more than 500 m from houses. Two approaches will help to make Tokat's green areas more planned and sustainable. The first suggestion is that the recommended housing zones should be relocated outside the buffer zone of current green areas, and the second is that a balance can be achieved by providing more green areas for the recommended housing zones.

Another approach is that recommended green areas should also complement one another and be unified. Green areas will thus be sustainable and form a climate that will help the urban ecology more. This will lead to more sustainable planning that will better reflect the city's urban structure.

4. Discussion

If green areas are not planned carefully in urban planning zones, they will have problems with accessibility and distribution. This problem of the planning means that the result of decision for increasing density of structure in the Tokat city center area is the reason of new additional structuring decisions made by changing relationship of the parcel and structure to attract and bring revenue to the area, which makes decreasing rate of open areas [16]. Plenty of green areas may not be an accurate reflection of realistic planning data. To plan green areas properly and healthily, it is clear that making decisions by considering the green areas will provide a great contribution to the planning approach. Green areas should be considered in planning not only from the necessity-based perspective but also from the perspective regarding the quality of providing ecological benefits, forming healthy living conditions and serving as the places of gathering and sheltering during disasters.

The first priority of the conservation planning is to protect historical buildings, and the second priority is the green area arrangement. However, it is well-known that green areas are so important in both improving the quality of life and increasing the visual quality. It is necessary to make mandatory planning arrangements for green areas with the regulations and laws. If it is not possible to do any planning action for green areas to serve conservation areas, at least the green areas should be created within 500-m distance rule as required by laws and regulation. This condition reveals the importance of green areas arrangements.

The analyses clearly showed that Tokat's green areas lacked connections and that their ecological aspects were not considered. Regarding the urban distribution of green areas which could be regarded as planned zones, 56% were outside the buffer zone, indicating that they have planning-related problems. This study shows the necessity to approach to green areas with a more planned and usable method.

5. Conclusion

This paper serves as an analysis guiding the efforts to study the accessibility of green areas in Turkey. Its results, which should interest Turkey, indicate that the green areas were designed without considering their accessibility, as the new regulation stipulates. Bringing green areas to within 500 m is important for the public and young, elderly, and disabled people, and this makes the ability to measure their accessibility important.

The results show that green areas are not designed in accessibility distance of 500 m. In addition, the buffer zone method gives us the opportunity to measure the accessibility of green areas. Green areas per person were taken into account for the urban planning in Tokat city. Among the obtained results, buffer zone has forest areas (61%) and woody areas (51%) which compensates the planning of the amount of green areas for the residential areas.

This study is intended to guide papers about the accessibility of urban green areas in Turkey. It proposes a method for measuring accessibility. The planning approaches seen in Tokat indicate the need for housing zones and green areas to be proportional and accessible, which is also the case for many other cities in Turkey. This study proves that ArcGIS is an effective program and an important tool for buffer zone measurement. This method is important in order to guide many future studies and to be used in new studies, for example, in urban fringe to create green zones, the efforts to create zones of approaching conservation areas in urban development, etc.

Acknowledgements

The author is responsible for planning especially about landscape architecture issue where urban conservation planning alteration is made in the parcel numbers 59, 132, and 391. The author developed the article for the whole city that included also these parcels. The author is so grateful to Tokat Municipality for helping to access data.

Author details

Mustafa Ergen
Faculty of Fine Art and Design, Department of Architecture, Siirt University, Siirt, Turkey

*Address all correspondence to: mustafaergen2002@yahoo.com

References

[1] Ergen YB. Investigation of Constitution Planning and Unplanning Development in Cities. Academic Works in the Field of Science, Culture and Art. 2018 Autumn. Ankara: Gece Publishing; 2018. pp. 143-156. ISBN: 978-605-288-877-3

[2] Morar T, Radoslav R, Spiridon LC, Pacurar L. Assessing pedestrian accessibility to green space using GIS. Journal of Transylvanian Review of Administrative Sciences. Romania; 2014;**42**:116-139. Available from: https://rtsa.ro/tras/index.php/tras/article/view/94 [Accessed: 23 April 2020]

[3] Comber A, Brunsdon C, Green E. Using a GIS-based network analysis to determine urban greenspace accessibility for different ethnic and religious groups. Landscape and Planning. 2008;**86**:103-114

[4] Ergen YB. Urbanism. Ankara: Higher Technical Teacher School; 1981. p. 353

[5] Natural England. Analysis of Accessible Natural Greenspace Provision Suffolk. 2010. Available from: publications.naturalengland.org.uk/file/6012346285686784 [Accessed: 23 April 2020]

[6] Koliotsis PT, Papadopoulou MP. The contribution of accessible urban greenspace in the quality of residents' life in the Attica basin—A spatial analysis. In: 15th International Conference on Environmental Science and Technology, Rhodes, Greece. 2017. Available from: https://cest2017.gnest.org/group/2261/proceedings.html [Accessed: 23 April 2020]

[7] IL MH. Design with Nature. New York: Wiley; 1969

[8] Hopkins L. Methods for generating land suitability maps: A comparative evaluation. Journal for American Institute of Planners. 1977;**34**(1):19-29

[9] Brail RK, Klosterman RE. Planning Support Systems. Redlands, CA: ESRI Press; 2001

[10] Collins MG, Steiner FR, Rushman MJ. Land-use suitability analysis in the United States: Historical development and promising technological achievements. Environmental Management. 2001;**28**(5):611-621

[11] Malczewski J. GIS-based land-use suitability analysis: A critical overview. Progress in Planning. 2004;**62**:3

[12] 2018 Environmental State Report for Tokat City. Tokat: Principle Branch for Environmental Impact Assessment (EIA) and Environmental Permission. 2019. Available from: https://webdosya.csb.gov.tr/db/ced/icerikler/tokat_2018_-cdr_rev-ze-20190807131602.pdf [Accessed: 23 April 2020]

[13] TURKSTAT. The Result of Population Registration System Based on Address. 2020. Available from: https://biruni.tuik.gov.tr/medas/?kn=95&locale=tr [Accessed: 23 April 2020]

[14] Ebregt A, De Greve P. Buffer Zones and their Management; Policy and Best Practices for Terrestrial Ecosystem in Developing Countries. Wateringen: JB&A Grafische Communicatie; 2000. Available from: https://edepot.wur.nl/118089 [Accessed: 23 April 2020]

[15] Ergen YB. The Framework of Planning Report for Urban Protected Area of Analytical Studies and Conservation Development. Tokat Conservation Development Plan Responsible Person Work in Gazi University in Revolving Fund

Management, Plan and Report Approval from Kayseri Cultural and Natural Heritage Borad in Ankara; 1992. pp. 1-102

[16] Ergen YB. . Investigation of Research for Negative Influence of Urban Development on Urban Protected Areas 2. Urban Conservation, Renewal and Applications Colloquium, Application Problems, Method, Techniques and Tools in Urban Conservation. İstanbul: Mimar Sinan University Architecture Faculty; 1994. pp. 85-94

Chapter 5

Disparity in Peri-Urbanisation Process in Lagos, Nigeria

Funmilayo M. Adedire

Abstract

This chapter assessed the causes of disparity in the peri-urbanisation process in Lagos new towns and the accompanying effect on the characteristics of the transitioning settlements. Data collection was primarily through administration of 384 and 370 questionnaires to purposively selected housing units in Ikorodu and Ibeju-Lekki, respectively. These two settlements represent the most rapidly urbanising peri-urban in Lagos State. Primary data collected included drivers of urban expansion, residents' demography, locational convenience and commuting frequency. To supplement the primary data, spatial images of 2006 and 2016 were acquired as satellite images from Google Earth archive for this study. Data analysis was carried out using descriptive statistics for the quantitative data and time series and satellite image analysis for the qualitative data. The results show a varying extent of transition primarily influenced by the residents' demography, linkages to the urban areas, quality of life and stakeholders' response to housing policy. The study concluded that urban policy should be used as a tool to ameliorate the disparity in infrastructure development, which is the major driver of changes, and also, government involvement in housing provision should have a spread in all urban periphery settlements in Lagos State.

Keywords: commuting, land use, peri-urbanisation, periphery, rural-urban linkages, socio-demography

1. Introduction

Peri-urbanisation in Lagos State is spearheaded by many actors often with conflicting interest [1]. The pattern and rate of land conversion from agricultural land use to residential and other land uses at the peripheral does not have corresponding infrastructure. Imbalance in changes in the peri-urban interface of most developing countries poses several challenges to planning and ultimately leads to a distortion in urban policy [2]. Urban transition in Lagos peri-urban interface is traceable to many factors, some of which include socio-demography, linkages to the metropolises, availability of basic services, availability and affordability of housing for the low income migrants and the state government policy on land use.

While the conversion of land at the peripheral of cities can enhance the provision of housing and space for industrial expansion [3], interactions of many forces at the peri-urban interface determine the sustainability. Primarily, the heterogeneous population makes it difficult for an effective community participation in the development of basic services [4, 5]. Urban quality of life is a function of participative planning and, this is greatly influenced by the socio-demography of the migrants which are mostly of the low income group in the periphery [6].

Urban form is a major determining factor in the extent and pattern of expansion. The dispersed nature of peri-urban settlements in most developing countries encourages government's disparity in infrastructure development. This is a major factor influencing the peri-urbanisation process in developing countries. Lack of urban compactness constitutes an economic burden in terms of providing basic services [7]. Isolated settlements from the urban core suffer from lack of infrastructure due to distance and the financial implications. Ease of commuting also encourages rapid expansion in cities' periphery [1]. Investigation by Adedire and Iweka [8] shows that effective linkages through different means of transportation are a catalyst in the development of metropolitan fringe. The expansion of settlements is in response to extent of linkages [9].

This study seeks to fill the gap in research on peri-urbanisation process in Lagos periphery. Some research works have been carried out on land use changes, rural-urban linkages and environmental quality in Nigerian peri-urban settlements [1, 10–12]. But none of these researches had assessed the disparity in the transition trends and the implications on the urban quality of life in Lagos peri-urban. Therefore, the aim of this study is to examine the causes of disparity in peri-urbanisation process in Lagos, Nigeria.

2. Literature review

One of the accompanying negative effects of urbanisation in developing countries is outward expansion of built up area and conversion of agricultural lands into residential and industrial uses [13]. The peri-urban expansion is part of a wider urbanisation process; it is the consequence of urbanisation [14]. Urbanisation in Lagos State, Nigeria has attached to it, peri-urbanisation which is an urban-driven transition in the territory outside the metropolitan regions or urban core [15].

The peri-urbanisation process is mostly in response to growth in population and spatial demand for industrial and economic activities. It is the gradual transformation of rural land to urban as a result of physical and human interactions [16]. The basic distinguishing factors between the rural, peri-urban and urban are population, built up density, infrastructure, administration, boundaries and economies [17]. Spatial morphology of rural-urban region includes the urban core, urban inner area, the suburban area, urban periphery, and rural hinterland [18].

Notable transformations in the peri-urban include socio-demographic, morphological, cultural, economic and functional changes [19]. The urban periphery in most developing countries, allows multi-dimensional development. It accommodates industrial activities due to the limited land area in the city centre and also provides a means of housing for the urban population [20]. In addition, peri-urban settlements also serve as a major provider of housing for urban population in highly populated cities in developing countries [7, 20, 21].

With high population growth rate and continuous rural-urban drift in most developing countries, housing deficit becomes a critical challenge [22]. This creates a seized opportunities in real estate development in creating well planned towns in the peri-urban. Therefore, peri-urban settlements in African nations are emerging with massive investment in housing development and infrastructure development because of the possibility of lower houses price of housing, large living space and a better environment. In addition, location of housing in the peri-urban is enhanced by transport, accessibility to employment and services. This is partially responsible for the growth of peri-urban settlements in Lagos [1, 8].

Drivers of housing expansion in the peri-urban are perceived infrastructural development, improved socio-economic activities and development of the tourism sector [23]. With improvement in transportation and the rising household incomes, which enhances more vehicle ownership, housing development in African peri-urban settlements is also increased [19]. Also the decline in urban environmental quality influences residential development in cities' periphery of developing countries [6]. Improved transport infrastructure has aided the spread of most peri-urban radially in all directions. Transport and communications are key factors in addition to infrastructure in facilitating and encouraging urban-rural migration regarded as counter urbanisation [10, 24].

In addition, the relocation of economic activity to peri-urban zones of most developing nations gradually leads to pull factors in terms of opportunity to engage in small industry, property investment, and improved urban quality, cheaper land for housing and improved infrastructure. Locational benefits are enabled by closeness to places of work, which is made achievable by improved transportation and the presence of public and private institutions, also commercial ventures [25].

The roles of different actors also inform the type and extent of urban transition in the peri-urban. Notable actors are local farmers, peri-urban residents, entrepreneurs, property developers and government institutions [26]. Entrepreneurs represent the demand side of the land market due to the spatial demand for commercial and industrial activities. The supply and demand of land market is controlled by property developers having greater hold on monetary and political powers, and in return influencing changes in the interface. They fill in the gap created by laxity of government in terms of formal control [26].

3. Study areas

To compare and contrast the peri-urbanisation process in Lagos State, Ikorodu and Ibeju-Lekki are selected as case studies. Ikorodu represents the highly urbanised peri-urban in Lagos in terms of residential development and population growth while Ibeju-Lekki represents the least urbanised peri-urban in Lagos in term of population growth and housing development. Ikorodu is located in the North East of Lagos State along the Lagos lagoon and situated at a distance of approximately 36 km north of Lagos. It occupies a land area of about 345 km^2. It is situated at approximately latitude 6°36′ North and longitude 3°30′ East. Ikorodu had an enumerated population of 535,619. The sample frame constitutes the existing buildings in the peri-urban. Ibeju-Lekki local government area is approximately 75 km long and

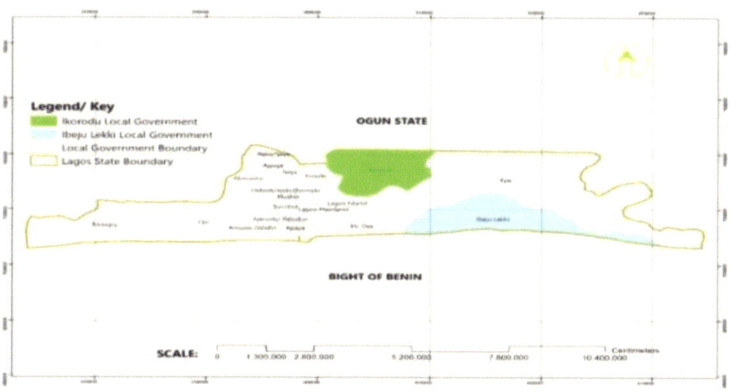

Figure 1.
Map of Lagos State showing the case studies. Source: Field work (2017).

about 20 km wide and has a land area of about 646 km^2, which equals one quarter of the total land mass of Lagos State. It is situated at approximately latitude 40°15′ North latitude 40°17′ North and longitude 13,015′ East and 13,020′ East. According to the National Population Commission [27] census, Ibeju-Lekki had a population of 117,481 out of Lagos State's total of 9,113,605 (**Figure 1**).

4. Methodology

Primary and secondary data for this study were extracted from the responses in the questionnaire, analysis of observation chart and the analysis of the spatial data. Quantitative data include the drivers of urban expansion, residents' demography, locational convenience and commuting frequency. Data analysis was carried out using descriptive method. Spatial data that is, satellite images were acquired from Google Earth while shape-files of the study areas were extracted from the state government's archive. Analogue spatial data were converted to digital and brought into ArcGIS environment in to reference them geographically.

5. Results and discussion

5.1 Land use changes between 2006 and 2016

It can be observed in **Figures 2a**, **b** and **3a**, **b** and that both Ikorodu and Ibeju-Lekki have experienced noticeable spatial expansion in terms of residential development between years 2006 and 2016. Ikorodu could be seen to have developed more in areal extent than Ibeju-Lekki as shown by **Figure 3** that greater percentage of the latter is still undeveloped. Ikorodu has less undeveloped area of land mass having experienced a surge in housing development within the study timeline. Urban expansion in both locations is motivated by different factors. Analysis of agricultural and residential land use in the study areas is presented in **Table 1**.

5.2 Chronological change in land use in the study area

Prior to 1980, peripheral settlements in the study areas were predominantly green areas used mainly for agriculture, conservation, water catchment, forest

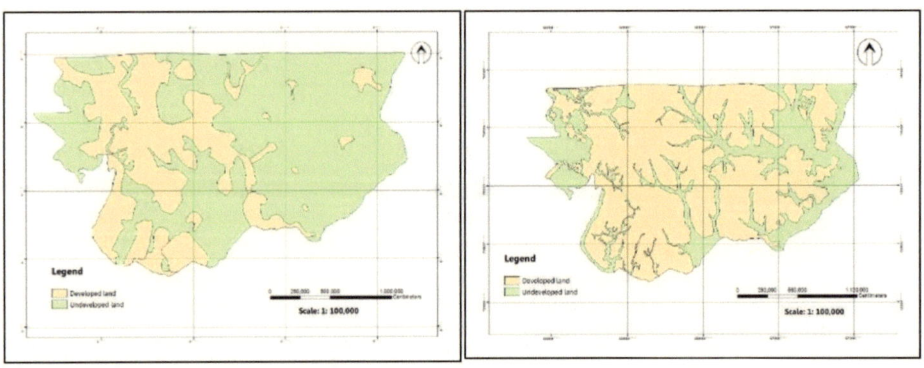

(a) (b)

Figure 2.
(a) Extent of residential expansion in 2006. (b) Extent of residential expansion in 2016. Development trend in Ikorodu study area 2006–2016. Source: Field work (2017).

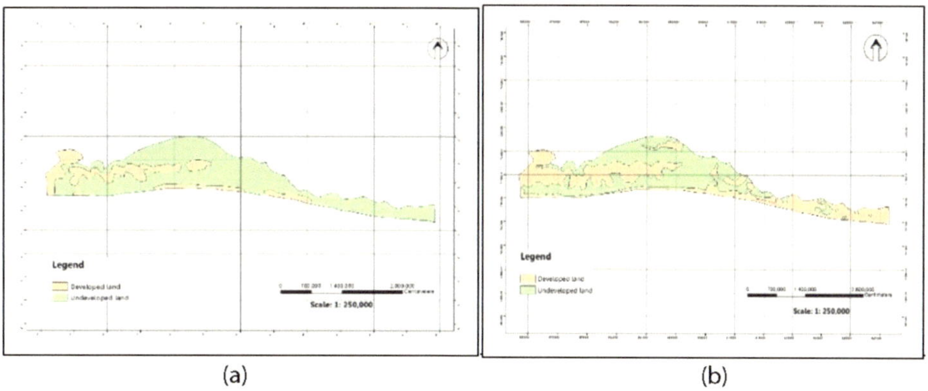

Figure 3.
(a) Extent of residential expansion in 2006. (b) Extent of residential expansion in 2016. Development trend in Ibeju-Lekki study area 2006–2016. Source: Field work (2017).

Land use	Pre-1980	1980–2000	2000	2002
		Ikorodu		
Agricultural	—	18,236.94	17,268.72	21,084.94
Residential	622.23	8799.78	12,492.33	17, 972.62
Other land uses	36,321.19	9906.70	7182.36	7885.85
		Ibeju-Lekki		
Agricultural	54,774.96	20,801.16	46,109.26	—
Residential	—	4049.27	6932.13	11,243.00
Other land uses	1830.06	31,751.60	3563.63	45,362.02

Source: Lagos State Ministry of Lands (2005).
'Other land uses: civic, commercial, business, conservation, water catchment, marshlands, fishing port, forest and wild palm trees, industrial, institutional, new ocean terminal, parks, open space, recreation and tourism, water body.

Table 1.
Pre 2006 agricultural and residential land use analysis in the study area.

and wild palm trees as shown by the analysis in **Table 1**. In Ikorodu, between years 1980 and 2000, agriculture became the major player in land use with gradual land conversion for residential purpose taking 8799.78 and 4049.27 hectares in Ikorodu and Ibeju-Lekki respectively. Residential land use continued to rise exponentially from 622.23 hectares in the year preceding 1980 to 17,972.62 hectares in 2002. Till date, agriculture and residential land uses have been the major forces behind the peri-urbanisation process in Ikorodu. The increase in residential land use in Ikorodu peri-urban is as a result of housing deficit in the neighbouring metropolitan area like Ikeja (the capital of the host state, Lagos). A gradual push into the peri-urban settlements by the saturation of core metropolitan Lagos due to rapid population growth and as a result of rural-urban drift led to residential development in all Lagos peri-urban settlements from year 1980.

The growth pattern in Ibeju-Lekki differs in terms of land use from that of Ikorodu. It has less land conversion. Also, residential land use in Ibeju-Lekki is not pronounced until 1980 when usage increased to 4049.27 hectares, and subsequently 6932.13 and 11,243.00 hectares in the year 2000 and year 2002 respectively. Peri-urbanisation in Ibeju-Lekki is a response to the development of the highways. Pioneering towns like Awoyaya and Abijo developed as a result of their closeness to the metropolitan centre of Lagos Island. Linear settlements grew along the major highways while the growth of the inner periphery was in response to land speculation. With land reclassification and government acquisition of land for development, infrastructure limited was necessary thus aiding the opening of the inner periphery for mixed use development. Residential segregation and residents' socio-demography influenced the settlement patterns. Various types of settlements also grew in response topography, culture and political reasons.

5.3 Residents' perception of urban expansion in the study area

From the perception of the residents, different factors were driving urban expansion in Ikorodu and Ibeju-Lekki periphery as analysed in **Table 2**. The major factors driving urban expansion in Ikorodu are low cost of living 20.8%, land affordability 18.2%, employment 16.1%, improved urban quality 12.9%, workplace location, and property investment. Ikorodu is known to house various institutions both private and public thus creating a haven for employment. In terms of land affordability, in comparison to the excessive cost of land in Lagos urban centre, Ikorodu provide good workplace and residential location at affordable cost. Other less important drivers are road accessibility, secured land tenure, cost of transportation, education and closeness to kinsmen. Peri-urbanisation process is driven by different principal factors in Ibeju-Lekki. In this periphery, major causes of the expansion in the study area are land affordability 27.3%, workplace location 14.8%, improved urban quality 15% and property investment 14.2%. Also vital to the growth of Ibeju-Lekki are employment and closeness to kinsmen. Government-led and private developer-led housing developments in the study area are on the increase because of the improved urban quality in Ibeju-Lekki.

5.4 Impact of residential land use on the study area

Data from the National Population Commission [27] census analysed in **Table 3** shows a huge leap in housing development in both Ikorodu and Ibeju-Lekki local government areas from the year 2006. Ikorodu local government has witnessed a more residential land conversion than Ibeju-Lekki. Housing developments in Ikorodu was 52,819 units in 2006, and 89,609 units in 2016

	Ikorodu		**Ibeju-Lekki**	
Drivers of expansion	**N = 379**	**%**	**N = 366**	**%**
Land affordability	69	18.2	100	27.3
Low cost of living	79	20.8	29	7.9
Workplace location	43	11.3	54	14.8
Improved urban quality	49	12.9	55	15
Employment	61	16.1	29	7.9
Study	7	1.8	3	0.8
Good and accessible road network	3	0.8	2	0.5
Low cost of transportation	10	2.6	1	0.3
Secured land tenure	14	3.7	15	4.1
Property investment	38	10	52	14.2
Closeness to kinsmen	16	4.2	26	7.1

Table 2.
Drivers of expansion in the study areas.

	Ikorodu	Ibeju-Lekki
Year	Housing units	Housing units
2006	*52,819	*7701
2016	**89,609	**11,749

Source: National population Commission [27].
**Source: Field survey (2017).*

Table 3.
Residential land use in the study area between years 2006 and 2016.

as obtained in the surface counting of selected 18 peri-urban settlements in Ikorodu. In Ibeju-Lekki local government, as noted by the National population Commission [27], a total of 7701 housing units were acknowledged in 2006 while during the surface counting of residential developments aided by aerial photographs during the field work in September, 2016, 11,746 units of housing were noted in the selected 16 peri-urban settlements in Ibeju-Lekki. In comparison to the urban core, individuals have access to affordable lands for housing and economic-related developments in the peri-urban and this is a major pull factor for peri-urbanisation in both cases.

5.5 Commuting patterns in the study areas

In the expansion of periphery settlements, linkages play a huge role. This comes in terms of interconnectivity between metropolises and the rural areas. Good commuting encourages a pull of urban population to the city periphery for residential and industrial development. The commuting patterns of the respondents in the case studies as presented in **Table 4** shows 38.8% of Ikorodu residents commute daily to the city, 36.1% on a weekly basis and 25.1% as the need arises. This reflects good level of linkage between Ikorodu periphery and the surrounding urban areas. The commuting hour shows 31.1% of the residents spend less than 30 minutes for daily commuting to work, 30.6% spend about 1 hour, 13.5% show a commuting time of 1 hour, 30 minutes, 16.6% spend between 1 and 2 hours, 5.8% spend 3 hours and other unspecified commuting time is about 1.6%. Residents' purpose of commuting to the city shows 54.4% of the respondents' population commute to the city for work, 23.2% for groceries, 20.3% for supply of their medium scale businesses and 0.6% for other needs.

In Ibeju-Lekki, 34.2% of the respondents' population travel to the urban centres daily, 33.3% commute to the city centre weekly and 31.4% travel as the needs arise. The highest commuting time to and from places of work daily in Ibeju-Lekki peri-urban is 3 hours while the least commuting time is 30 minutes. 35.2% spends an average of 60 minutes (1 hour) daily commuting, 24.6% spend 90 minutes, 15.6% spends less than 30 minutes, 14.2% spends almost 180 minutes (3 hours) while 10.4% spend an average of 120 minutes (2 hours) commuting daily. 42.6% of the respondents travel for work related purpose, 30.3% travel to either the city centre and neighbouring peri-urban for groceries while 27% travel to the city centre for supply of materials for their enterprises.

5.6 Impact of locational convenience on spatial expansion of the study area

Convenience is vital in the expansion of cities 'periphery'. Factors like closeness to work, closeness to market or the central business district, and availability of public transport were considered in the study areas.

	Ikorodu		Ibeju-Lekki	
	N = 379	%	N = 366	%
Frequency of commuting				
Daily	147	39	125	34.2
Weekly	137	36	122	33.3
Others (specify)	95	25	116	31.7
Average time of commuting				
Less than 30 minutes	118	31	57	15.6
31–60 minutes (1 hour)	116	31	129	35.2
61–90 minutes (1 and half hours)	51	14	90	24.6
91–120 minutes (2 hours)	63	17	38	10.4
121–180 minutes (3 hours)	22	5.8	51	13.9
Unknown	9	2.4	1	0.3
Purpose of commuting				
Work	206	54	156	42.6
Groceries	88	23	111	30.3
Supply for business	77	20	99	27
Others	8	2.2	0	0

Source: Field survey (2017).

Table 4.
Commuting patterns in the study areas.

5.6.1 Availability of public transport

Good public transportation system exists in Ikorodu as shown in **Table 5** and **Figure 4**. Among the respondents, 83.9% have access to good transport system while 15.6% do not. In Ibeju-Lekki, 22.1% of household heads lack good public transportation due to their residential location in the inner periphery. This is translated to high cost of transportation for residents without personal vehicles in such areas as shown by observation during the field survey. However, 77.6% indicated good public transportation at their disposal.

5.6.2 Closeness to work

In Ikorodu, good accessibility to work is enhanced by the strategic locations of most institutions and the efficient distribution of government organisations around both primary and secondary roads. 72.8% of the respondents' population were privileged to have good proximity to their works while 26.9% were not as presented in **Table 5**. In Ibeju-Lekki, a greater percentage of the respondents have good workplace location than in Ikorodu. 85.8% indicated locational benefits while 14.2% are were negatively affected by their residential location in relation to places of work. These findings gives strength to Alonso's access trade off model which states that the choice of residential location in periphery settlements could offer an opportunity lost and opportunity cost. A trade-off exists between cheaper land in the peri-urban and the hours of commuting (**Figure 5**).

	Ikorodu		Ibeju-Lekki	
Variable	**N = 379**	**%**	**N = 366**	**%**
Availability of public transport				
Yes	318	83.9	284	77.6
No	59	15.6	81	22.1
Unknown	2	0.5	1	0.3
Closeness to work				
Yes	276	72.8	314	85.8
No	102	26.9	52	14.2
Unknown	1	0.3	0	0
Closeness to CBD				
Yes	273	72	315	86.1
No	104	27.4	51	13.9
Unknown	2	0.5	0	0

Source: Field survey (2017).

Table 5.
Analysis of locational convenience in the case studies.

Figure 4.
Road networks and transportation system in Ikorodu periphery. Source: Field survey (2017).

Figure 5.
Workplace locations in Ikorodu periphery. Source: Field survey (2017).

5.6.3 Closeness to the central business district

In Ikorodu periphery, 72.0%, a reasonable size of the respondents had residential location advantage by virtue of their proximity to the central business districts, but 27.4% were disadvantaged. 86.1% of respondents in Ibeju-Lekki periphery were close to the central business district while 13.9% were not close to the central business district leading to increased commuting to the city.

6. Conclusion and recommendation

There have been phenomenal changes in land use and land cover of Nigerian peripheral interface. New towns outside the metropolises of developing countries undergo expansion due to population growth and housing deficits in the urban areas. The spatial transformation is mostly accompanied with encroachment into the surrounding agricultural land. The study areas in this study are experiencing conversion in land use but differ in pattern and extent. Findings show generally that the major factors influencing peri-urbanisation are land affordability, improved road network and the emergence of gated exclusive housing developments by private developers in the peri-urban areas.

There exist disparity in the peri-urbanisation process in the study areas and this could be attributed to many factors, among which are the level of the linkages, locational convenience (both for workplace and residential development), people's socio-demography and the government housing policy. The degree of establishment of each of these variables in the study areas determines the rapidity of the peri-urbanisation process. The wider land conversion experienced in Ikorodu is traceable to the presence of these variables in the periphery.

Availability of good transportation network is a motivation for living in the suburb. Locational benefit is a huge pull factor to the peripheral for residential developments. The commuting pattern shows the interdependencies between the periphery and urban centres, mostly for socio-economic purposes. This corroborates the findings of [1]. Most peri-urban residents work in the city centre while having residential location in the metropolitan fringe. Both locations however have good proximity to the Central Business District. Also the ability of housing providers to put locational convenience into consideration would promote balanced development.

Linkages show that the peri-urban cannot function in isolation. The first case study, Ikorodu has been subjected to rapid changes due to locational convenience and availability of public transportation. The frequency of commuting in Ikorodu is more than that of Ibeju-Lekki. Findings show work-related activities as the major reason for commuting to the urban centre in Ikorodu peri-urban. Commuting time from the peri-urban to the city centre for various activities is longer in Ibeju-Lekki than Ikorodu and this could be attributed to the nature of the settlements that are mostly dispersed.

Also improved infrastructure and concentrated development of institutions in Ikorodu are major contributors to the rapid expansion of the peri-urban interface. The peri-urbanisation process is usually slowed down where there is dearth of infrastructure and socio-economic institutions. It can be concluded that equality in infrastructure development will limit disparity in peri-urbanisation process. Prior studies by Binns et al. [11] and Lawanson et al. [1] have shown that Lagos as a rapidly urbanising region is not exempted from the peri-urbanisation phenomenon.

Author details

Funmilayo M. Adedire
Department of Architecture, Lead City University, Ibadan, Oyo State, Nigeria

*Address all correspondence to: funmidire@gmail.com

References

[1] Lawanson T, Yadua O, Salako I. An investigation of rural-urban linkages of the Lagos megacity, Nigeria. Journal of Construction Project Management and Innovation. 2012;**2**(2):464-581

[2] Allen A. Environmental planning and management of the peri-urban interface: Perspectives on an emerging field. Environment and Urbanization. 2003;**15**(1):135-147

[3] Wu F, Zhang F, Webster C. Informality and the development and demolition of urban villages in the Chinese peri-urban area. Urban Studies. 2013;**50**(10):1919-1934

[4] Allen A. Neither rural nor urban: Service delivery options that work for the peri-urban poor. In: Peri-Urban Water and Sanitation Services. Netherlands: Springer Netherlands; 2010. pp. 27-61

[5] Obeng W, Whittal J. Peri-Urban Infrastructure Development through Community Participation: A Case Study of Yasore, Ghana. ResearchGate. 2014. pp. 1-13

[6] Adedire F, Adegbile M. Effects of urbanization on spatial demography change in Ibeju-Lekki, Lagos peri-urban settlement. Community Development. 2018;**49**(3):292-311

[7] Browder JO, Bohland JR, Scarpaci JL. Patterns of development on the metropolitan fringe: Urban fringe expansion in Bangkok, Jakarta, and Santiago. Journal of the American Planning Association. 1995;**61**(3):310-327

[8] Adedire F, Iweka A. Implications of residential and workplace location in Ikorodu. Ethiopian Journal of Environmental Studies and Management. 2017;**10**(8):1082-1093

[9] Adedire F. Peri-urban expansion in Ikorodu, Lagos: Extent, causes, effects and policy response. Urban Forum. 2018;**29**(3):259-275

[10] Bah M, Cissé S, Diyamett B, Diallo G, Lerise F, Okali D, et al. Changing rural–urban linkages in Mali, Nigeria and Tanzania. Environment and Urbanization. 2003;**15**(1):1-12

[11] Binns J, Maconachie R, Tanko A. Water, land and health in urban and peri-urban food production: The case of Kano, Nigeria. Land Degradation & Development. 2003;**14**(5):431-444

[12] Nwokoro II, Dekolo SO. Land use change and environmental sustainability: The case of Lagos Metropolis. In: The Sustainable City VII: Urban Regeneration and Sustainability. RSA: WIT Press; 2012. pp. 157-167

[13] Satterthwaite D. Urbanisation in low and middle income nations in Africa, Asia, and Latin America. In: Satterthwaite D, editor. The Companion to Development Studies. London: Routledge; 2014. p. 279

[14] Allen A, Da Silva NL, Corubolo E. Environmental problems and opportunities of the peri-urban interface and their impact upon the poor. In: Strategic Environmental Planning and Management for the Peri-Urban Interface Project. London: University College London; 1999

[15] Dekolo S, Oduwaye A. Managing the Lagos megacity and its geospatial imperative. In: 28th Urban Data Management Symposium (UDMS 2011); Delft, The Netherlands. International Archives of the Photogrammetry, Remote Sensing and Spatial Information Sciences. Vol. XXXVIII-4/C21. 2011. pp. 28-30

[16] Adell G. Theories and models of the peri-urban interface: A changing conceptual landscape. In: Development Planning Unit. London: University College London; 1999

[17] Ravetz J, Fertner C, Nielsen TS. The dynamics of peri-urbanization. In: Nilsson K, Al E, Nilsson KP, editors. Peri-Urban Futures: Scenarios and Modals for Land Use Change in Europe. Netherlands: Springer-Verlag Berlin Heidelberg; 2013. pp. 13-44

[18] PLUREL. Peri-Urban Transition Processes. Denmark: PLUREL; 2009

[19] Dutta V. Land use dynamics and peri-urban growth characteristics: Reflections on master plan and urban suitability from a sprawling north Indian city. Environment and Urbanization ASIA. 2012;**3**(2):277-301

[20] Mcgregor D, Simon D, Kwasi N G. The changing urban–rural Interface of African cities: Definitional issues and an application to Kumasi, Ghana. Environment and Urbanization. 2004;**16**(2):235-248

[21] Acheampong RA, Anokye PA. Understanding households' residential location choice in Kumasi's peri-urban settlements and the implications for sustainable urban growth. Research on Humanities and Social Sciences. 2013;**3**(9):60-70

[22] Mcgranaham G, Satterthwaite D. Urbanisation Concepts and Trends. London: Institute for International Economics and Development; 2014

[23] Thapa RB, Murayama Y. Drivers of urban growth in the Kathmandu Valley, Nepal: Examining the efficacy of the analytic hierarchy process. Applied Geography. 2010;**30**(1):70-83

[24] Rozmi I, Mohammed H, Rahim MN. Passengers preferences and satisfaction of public transport in Malaysia, Part II: A comparative analysis of Komuter and LRT network. Research Journal of Applied Sciences, Engineering and Technology. 2013;**6**(8):1450-1456

[25] Pradoto W. Development Patterns and Socioeconomic Transformation in Peri-Urban Area. Berlin: Univerlagtuberlin; 2012

[26] Salem M. Peri-urban dynamics and land-use planning for the greater Cairo region in Egypt. Sustainable Development. 2015;**1**:109-119

[27] National Population Commission. National Census and Demography. Federal Republic of Nigeria: National Population Commission; 2006

Section 3

Landscape Sustainability: Methods and Practices

Chapter 6

Landscape Hazards: Destructive Build Environment Zones and Safe Areas - An American Case Study

Yoichi Kunii, Paige O'Keefe, Jon Burley, Luis Loures and Marifaye Regina Villanueva

Abstract

Planners, designers, governmental organizations, and citizens are interested in creating enduring safe buildable environments. Landscape hazards such as earthquakes, wildfires, hurricanes, tornados, flooding, volcanoes, radon, air pollution, sinkholes, avalanche, landslides, and blizzards create a complex set of destructive forces that form disturbances obliterating life and structures. In our study, we examined these forces across the lower 48 states of the United States of America. We applied geographic information system (GIS) technology to identify areas of extreme hazard and areas of low risk. Our investigation indicated that most of our study area (approximately 83%) was exposed to highly reoccurring destructive forces and that only relatively small patches (Upper Midwest-portions of Michigan, Wisconsin, and Minnesota) and thin stretches (Rocky Mountain Front Range—eastern Montana, Wyoming, and eastern Colorado) of land were relatively secure from these forces. This means that in the long term, much of the study area is not safe from disturbances that will destroy much of the built environment, challenging notions of sustainability for numerous metropolitan areas, United Nations Educational, Scientific, and Cultural Organization (UNESCO) Biosphere Reserves, UNESCO World Heritage Sites, National Parks, other noted historic sites.

Keywords: environmental geology, environmental planning, landscape architecture, natural resources, physical geography

1. Introduction

Safe, enduring, sustainable built environments are of great interest to planners, designers, governmental organizations, and citizens. Yet yearly across the globe, built environments are destroyed by tsunamis, hurricanes, earthquakes, wildfires, tornadoes, volcanoes, flooding, landslides, avalanches, and other environmental hazards. The loss of life and damage to property is extensive. As each event occurs, scholars study the cause of the event, the extent of the damage, and impact upon the environment. For example, Foxworthy and Hill describe the cataclysmic event of the Mount St. Helens volcanic eruption of 1980—this event was only a relatively small volcanic eruption [1]. Ekey recounts the extent and damage of the 1988 Yellowstone fire; while Daniel and Ferguson edited a series of papers discussing the knowledge concerning wildfires and the urban interface [2, 3]. Stanley Changnon edited a document describing the extensive flooding event in the Mississippi River Basin of 1993 [4]. Numerous authors describe earthquake events ranging from events in relative wilderness to urban areas [5–9]. Margot Keam Cleary describes many more events of the twentieth century, noting avalanches, hurricanes/typhoons/cyclones, tornados, and tsunami/tidal waves [10]. In addition, authors have described catastrophic events such a meteorite collisions and atmospheric poisoning leading to changes in the composition and structure of the biosphere [11]. Each event would raise public awareness, but for many in the planning and design community, environmental hazards and the long term suitability of a building site were of minor importance when compared to issue of landscape conservation, design beauty, economics, and short term functionality [12, 13]. To illustrate this perspective, in the United States of America, Falling Water/Kaufman House, design by the acclaimed American architect of the twentieth century Frank Lloyd Wright in about 1935 is considered to be one of the great pieces of architecture for that century; yet in a 100 year flood, the waters of the seemingly serene creek rise to the mid-level of the living room (**Figure 1**) [11].

By the 1960s, planners and designers in the United States of America explored approaches to place built environment facilities in safe zones compatible with the structural ecology of the area, as illustrated by the barrier islands study of Ian McHarg and placing structures outside the path of avalanche zones at Snowbird, Utah by Dan Kiley [14, 15]. Landscape architects had expanded their work to encompass landscape planning studies, something that had not been widely practiced since efforts earlier in the twentieth century by Warren Manning [11]. For example, the complete land area and some aquatic habitats of the state of Hawaii have been completely planned and zoned with assistance of the professional design firm EDAW, led by Garrett Eckbo (the "E" in EDAW). The landscape is divided into areas for housing, recreation, grazing, crop production, forestry, armed services usage, conservation, and for use by the native Hawaiian people. The plan included considerations for mitigating the effects of three natural hazards: earthquakes,

Figure 1.
The red line approximates the level of the 100 year floodplain at the Falling Water House in Pennsylvania, USA (copyright © 2007 Jon Bryan Burley, all rights reserved, used by permission). In flooding conditions, the structure even acts as an obstruction to water flow, something that is often now prohibited for many areas of the United States.

tsunamis, and volcanoes [16, 17]. This general approach was applied by Burley and Burley to a study site in Colorado, to determine safe building environments against wildlife, avalanche, rock fall, and flooding. They determined that in their study area, there was no safe site [18]. This interest extended to other areas in the world, as Feng et al. examined building site safety in the Wenchuan are of China and in the central Philippines in post-earthquake settings developing an index to assess and determine the resiliency of the setting to save lives [19]. But in many respects, response to landscape hazards in planning and design had been practiced by some in other parts of the world, long before Americans began to study such topics. For example, in Tokyo, Japan, the Kiyosumi Garden, developed in 1878–1885 was created as a safe-haven in post-earthquake events and together with a nearby public park, remains as a post-earthquake safe-haven and was used as a safe haven during the allied/American bombings of Tokyo in 1945, **Figure 2** [20]. Similar work concerning safe haven open space has been recently studied in the Chinese province of Fujian [21–23]. These examples illustrate that at times investigators, public officials, and concerned citizens have occasionally/sporadically addressed hazards in the built environment; however, interest in this topic has increased.

Community resilience is an increasingly addressed issue worldwide, as it encompasses a widespread usage of resources by community members that allows them to thrive in a constant state of change and unpredictability [24]. As climate change develops into an increasingly more harmful and destructive force, communities need to be able to withstand and recover from these devastating effects. Presented as an opportunity to face vulnerability with resilience, climate change is the quintessential factor which immediately is threatening both our natural and human systems. The need to establish, enhance and promote tools for the overall health and safety of communities is increasing; thus, Community Resilience Assessment (CRA) tools have continued to evolve over the course of the twenty-first century [25]. Resilience, a term consisting of varying definitions, is composed of the same underlying concept of a mix of natural and mechanical systems with the ability to adapt to extreme shocks and uncertainty [26]. When creating and planning a design, policy makers, developers, landscape architects, and other professionals involved in the process, all play a vital role in the implementation of community resilience. Although many are currently aware of the effects and future possibilities environmental change and landscape hazards, there are also many who are not thinking about the essential planning steps needed to be able to withstand these effects.

In an attempt to depict a dynamic system responding to hazards and change that is not necessarily in balance, Graham A. Tobin cohesively created a conceptual framework for analysis of sustainability and resilience that consists of three separate

Figure 2.
A view of the Kiyosumi Garden in Tokyo, Japan is an open space that remains as a refuge for post-earthquake events (copyright © 2019 Jon Bryan Burley, all rights reserved, used by permission).

heuristic normative theories (meaning based upon expert opinion): a mitigation model, recovery model, and structural-cognitive model [27]. Collectively, these provide an in-depth look at the realities of implementing a sustainable and resilient framework that demonstrates the difficulties such as local context, social and political activities, and economic concerns. Tobin's ideas have been adapted from the works of Waugh and Mazmanian and Sabatier [28, 29]. In order to prevent high levels of exposure and risk, acts of prevention are critical to a community's success in the complete cycle hazard recovery and resilience. An example given are the mitigation policies that ensure specific conditions are met when implementing design standards of flood embankments and levee systems. Thus, a physical action is being taken towards the overall community resilience instead of the issue remaining theoretical which does not provide any measurable outcome. These conditions were then condensed into six major priorities for successful implementation: (1) sound theory with causal linkages to assure reasonable goals; (2) tasks and programs must be assigned to sympathetic agencies with adequate resources; (3) leaders must have managerial and political skills; (4) clear policy objectives with long term commitments; (5) organized constituency support; (6) no undermining of the policy over time [27]. Overall, these conditions and goals must be clearly articulated in order to provide safety, resilience, and resources over time to a wide variety of communities.

The state-of-the art concerning landscape hazards suggests that there is a wider concern across the public and professional ability and interest in assessing and implementing plans and design related to this issue. Still, the effort is case by case, city by city, and region by region. Rarely has there been an examination of a broad set of hazards for a substantially large area. Reporting of hazard events is often in the national and international news cycle. As this article is being completed, the wildfires of Australia are in the news [30]. In some respects there is no comprehensive study because no governmental agency is fully/completely

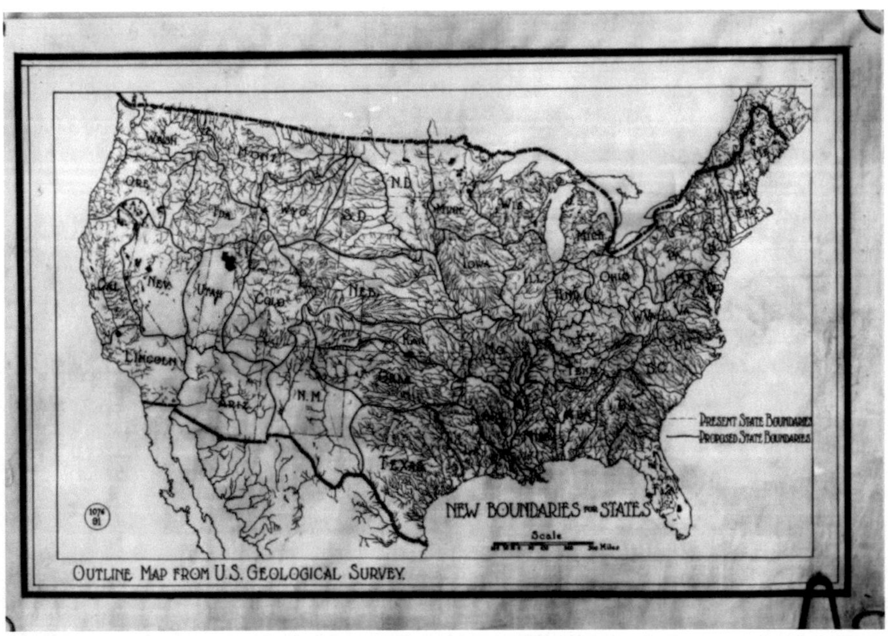

Figure 3.
A page from Warren Manning's National Plan, with a reorganization of the American states based upon physical/watershed boundaries (copyright © expired, obtained from the Iowa State University Library Special Collections and University Archives) [31].

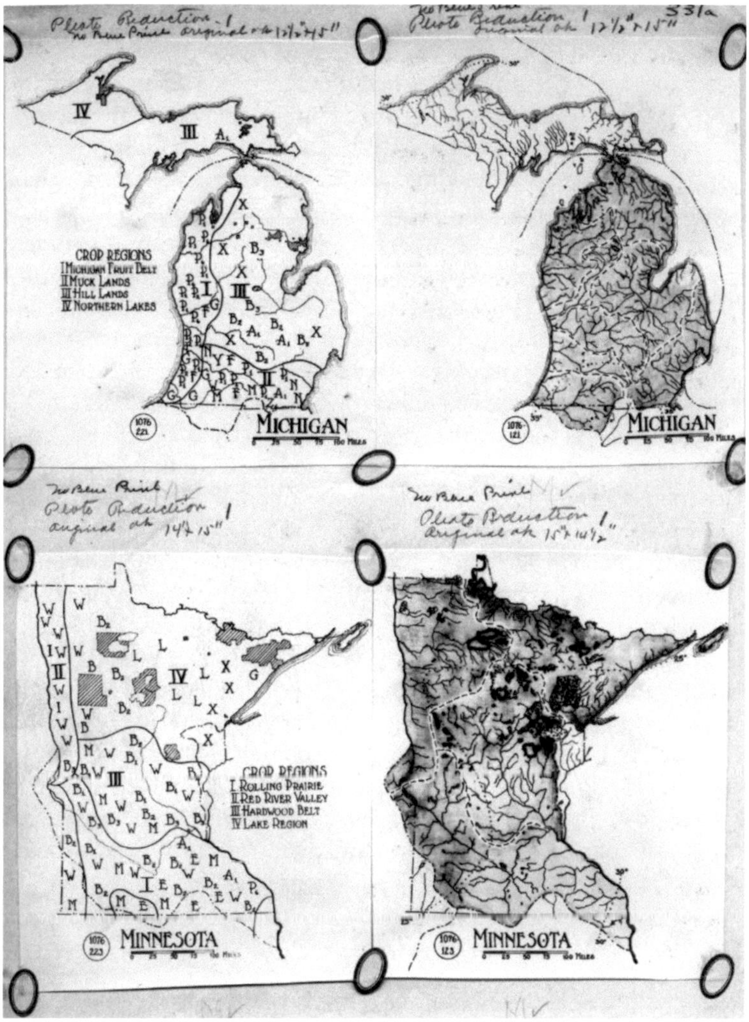

Figure 4.
A page from Warren Manning's National Plan, illustrating how the new states of Michigan and Minnesota are divided into management regions for agriculture, forestry, and conservation (copyright © expired, obtained from the Iowa State University Library Special Collections and University Archives) [31].

responsible to address planning and design for all types of hazards (the most comprehensive agency responds to hazards in the United States, the Federal Emergency Management Agency (FEMA) advising the public concern advanced preparation for some types of hazards such as earthquakes, wildfires, tornados, and hurricanes). Unfortunately no investigatory team has been funded to examine this issue in the same manner as Warren Manning, who lived from 1860 to 1838, who prepared a national comprehensive conservation management plan for the United States (**Figures 3** and **4**) [31]. He did not examine landscape hazards. But if he was living today, maybe it would be an issue that he might address.

We wondered if it was possible to address the lower 48 states concerning a multiplicity of landscape hazards to gain a more comprehensive understanding of the issues facing the built environment and long-term sustainability of building sites? In our investigation we were curious about: are there only small areas that merit hazard planning and design?; are there numerous and extensive areas that are relatively safe zones?; and what is the situation in the lower 48 states?

2. Methodology

To conduct the study, the team examined the same basic setting as Warren Manning [31]. The investigatory team gathered public data concerning a set of landscape hazards across the lower 48 of the United States, including: earthquakes, wildfires, hurricanes, tornados, flooding, volcanoes, radon, air pollution, avalanche, landslides, sinkholes, and blizzards [32–40]. The maps were drawn in layers with three values: high risk (medium gray with a 10–200 year time frame), moderate risk (light gray 500 year time frame), and low risk (white great than 500 year time frame), similar to Burley and Burley [18]and McHarg [14]. The model to compile the maps in a series of overlays was similar to Johnson and Burley, where the most hazardous value (a medium gray) across the overlays determined the hazard risk for a location [41]. Only locations with no high (medium gray) or moderate hazard rating (light gray) would receive a low (near white) hazard rating [41]. Locations with no value in the hazardous rating and with a maximum of a moderate rating would appear in the results map a moderate rating. For example, a site with a moderate earthquake score and all other scores being low, would be rated as a moderate (light gray) hazardous area. No effort was made to derive weighted maps or maps with linear combinations. As of yet, no investigator has demonstrated that the hazard layers should be combine in some latent dimension or equation. Although in the future, investigators might explore statistical relationships amongst the variables, as other investigators had done in visual quality and soil reclamation studies [42, 43]. The late Phil Lewis did discover that wetlands, slopes that require protection, and recreational lands covaried forming corridors, suggesting a latent dimension in environmental conservation and recreation to for greenways [44]. But so far, no such work has been accomplished with hazard data. In this hazard study the resulting map in this investigation may appear with many levels of gray (darker indicate many hazards and white indicating no hazards).

3. Results

The resultant map (**Figure 5**) contained approximately 83% of the study area with high and moderate hazard ratings. The locations with a fair expanse of low ratings occurred in the rain shadow (east) of the Rocky Mountains on the western

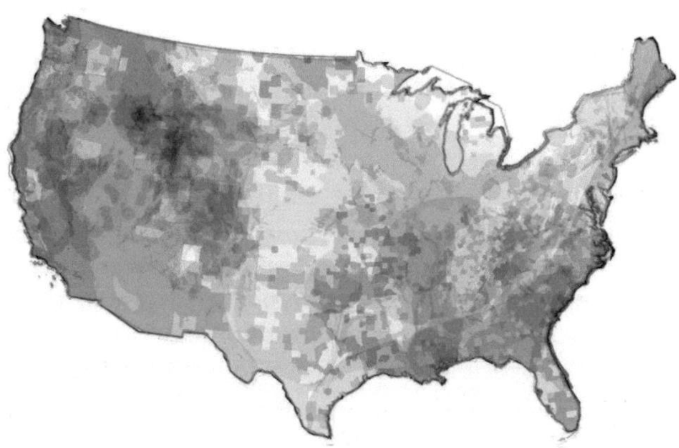

Figure 5.
A map of the hazard areas in the lower 48 of the United States when all the hazard maps are combined together (copyright © 2018 Yoichi Kunii and Jon Bryan Burley, all rights reserved, used by permission).

edge of the Northern Great Plains from west Texas to Montana and a smaller swath of land in the upper Midwest (Michigan, Wisconsin, northern Minnesota). A patchwork of lighter gray also occurs on the west side of the Appalachian Mountain in the Tennessee and Ohio River valleys north towards Pennsylvania and New York. However, there is no truly completely safe site. Smaller, county sized patches of relatively low hazard areas occur in the mountain west.

4. Discussion

When the environmental hazards are combined together, it become clear that much of the landscape will encounter some sort of hazard that may affect the built environment. The map suggests that over a 200 year period (10 generations), most sites will encounter some sort of hazard. While for any one generation, a group of individuals or community may experience no hazard event, in the higher hazard areas, events may be frequent across generations. The map in **Figure 5** indicates that much of the country will face repeated events and that there are relatively few refuges. This may be a surprise to some citizens and public officials who may expect their environments to remain stable and safe long term. The map suggests that building sites may be disturbed, even destroyed at a frequent rate, meaning within 10 generations. The disturbance probability is much greater than for just some unlucky locations such as in the San Francisco area, the gulf coast in the south east, or in and near Yellowstone National Park.

What does this mean for the built environment? For long term sustainability, care and thought may have to be given to mitigating the expected forthcoming event. Building codes and site design may have to reflect minimizing damage and sustaining life.

Figure 6 presents a map containing United Nations Educational, Scientific and Cultural Organization (UNESCO) Biosphere and Cultural Heritage areas, plus National Parks, and other historic landscape architectural sites described by Newton, Tobey, and Burley and Machemer [11–13].

The map illustrated in **Figure 6** suggests that many valued natural environments, cultural sites and other valued landscapes are in zones that will be exposed

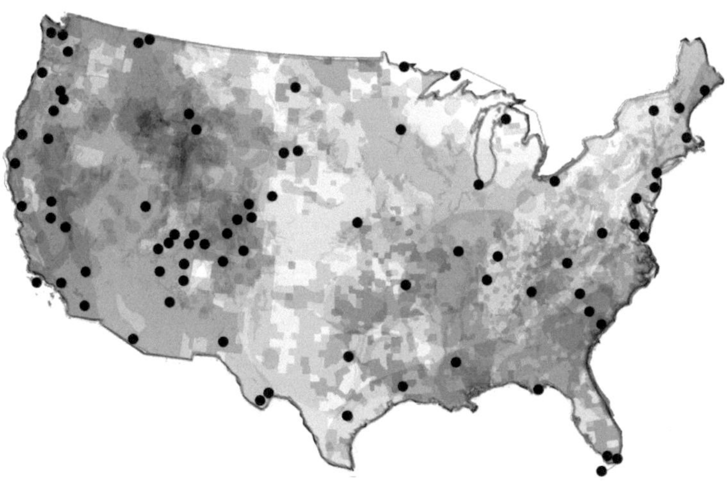

Figure 6.
A map illustrating the locations of valued landscapes across the landscape hazards composite map are combined together (copyright © 2019 Yoichi Kunii and Jon Bryan Burley, all rights reserved, used by permission).

to disturbance. Only a few sites on the Great Plains or in northern Michigan and Minnesota may be in areas with little change from hazards. Change is coming. Often individuals may assume that these sites may remain undisturbed and unaffected for many centuries. But the truth may be that many of these sites will encounter events much sooner than expected. Very few sites may have the longevity that the Pyramids of Giza in Egypt have endured. After all, the other six wonders of the world are in ruins [11]. Even places like central Michigan exposed to few events, over the last 12,000 years endured mile high glaciers, large fluctuations in the level of the Great Lakes, the extinctions of mammoths (*Mammuthus primigenius* (Blumenbach, 1799 [originally *Elephas*])) and mastodons (*Mammut americanum* Kerr 1792), the migration of vegetation from the south, the clearing for forests, the coming of urbanization, the automobile, and the invasion of exotic species [11].

In this study, there are more variables that could be included, such as water or soil pollution, or the impacts of various climate change scenarios. In addition, it could be debated about how the variables were classified and combined, or possibly a different base map for a certain variable could be used. Other investigators could generate variations on the results. This study is not definitive.

The environmental dangers to building sites are real and extend to nations around the world (**Figure 7**). The recent eruption of the Taal volcano in the Philippines illustrates the dangers to the built environment as it is an earthquake zone and volcano hazard area [45]. This is the same area that was hit by Typhoon Phanfone (Ursula) in late December 2019 [46].

Figure 7.
An image of the Taal Volcano erupting in January 2020 as seen from Los Baños, Philippines. The volcano is erupting tens of kilometers away, beyond the mountain/hills in the back of the image (copyright © 2020 Marifaye Regina Villanueva, all rights reserved, used by permission).

5. Conclusion

Planners and designers are engaging issues related to examining larger landscapes. This engagement facilitates understanding factors, forces, and influences upon the built environment. In this investigation, it was discovered that much of the study area will experience hazards events that will perturb the built environment, sooner than some might expect. To be sustainable or resilient may mean that these disturbances may require thoughtful adjustment by citizens, government officials, the construction industry, and planning/design professionals. Landscape architecture has become a profession engaged in examining broader environmental concerns beyond site planning and detailed design.

Conflict of interest

The authors declare no conflict of interest.

Author details

Yoichi Kunii[1], Paige O'Keefe[2], Jon Burley[2]*, Luis Loures[3] and Marifaye Regina Villanueva[4]

1 Tokyo University of Agriculture, Tokyo, Japan

2 Michigan State University, E. Lansing, MI, USA

3 Instituto Politécnico de Portalegre, Portalegre, Portugal

4 Independent Scientist, Los Baños, Philippines

*Address all correspondence to: burleyj@msu.edu

References

[1] Foxworthy BL, Hill M. Volcanic Eruptions of 1980 at Mount St. Helens the First 100 Days. Washington, DC: US Department of the Interior; 1982. Geological Survey 1249

[2] Ekey R. Yellowstone on Fire! Billings. Montana, USA: Billings Gazette; 1989

[3] Daniel TC, Ferguson IS, editors. Integrating research on hazards in fire-prone environments: Reconciling biological and human values in the wildland/urban interface. In: Proceedings of the U.S.—Australia Workshop. Melbourne, Australia: The United States Man and the Biosphere Program; 1989

[4] Changnon SA, editor. The Great Flood of 1993: Causes, Impacts, and Responses. Boulder, Colorado, USA: Westview Press; 1996

[5] Christopherson E. The Night the Mountain Fell: The Story of the Montana-Yellowstone Earthquake. West Yellowstone, Montana, USA: Yellowstone Publications; 1962

[6] Atkinson W. The Next New Madrid Earthquake: A Survival Guide for the Midwest. Carbondale, Illinois, USA: Southern Illinois University Press; 1989

[7] Cohen S. 8.6 the Great Alaska Earthquake March 27, 1964. Missoula, Montana, USA: Pictorial Histories Publishing CO., Inc.; 1995

[8] McNutt SR, Snydor RH, editors. The Loma Prieta (Santa Cruz Mountains), California, Earthquake of 17 October 1989. Sacramento, California, USA: Department of Conservation, Division of Mines and Geology; 1990

[9] Bronson W. The Earth Shook, the Sky Burned: A Moving Record of America's Great Earthquake and Fire: San Francisco, April 18, 1906. San Francisco, California, USA: Chronicle Books; 1986

[10] Cleary MK. Great Disasters of the 20th Century. New York, New York: Gallery Books; 1990

[11] Burley JB. Machemer From Eye to Heart: Exterior Spaces Explored and Explained. San Diego, California, USA: Cognella Academic Publishing; 2016

[12] Newton NT. Design on the Land: The Development of Landscape Architecture. Cambridge, Massachusetts, USA: The Belknap Press; 1974

[13] Tobey GB Jr. A History of Landscape Architecture: The Relationship of People to Environment. New York: American Elsevier Publishing Company, Inc.; 1973

[14] McHarg IL. Design with Nature. Garden City, New York: Doubleday/Natural History Press; 1969

[15] Yamada M. Dan Kiley: Landscape Design Two: In Step with Nature. Tokyo: Books Nippon; 1993. Process Architecture Series: No. 108

[16] Eckbo D. Austin & Williams. State of Hawaii Land Use Districts and Regulations Review. Honolulu, Hawaii, USA: State of Hawaii Land Use Commission; 1969

[17] Armstrong RW, editor. Atlas of Hawaii. Honolulu. Hawaii, USA: The University of Hawaii Press; 1973

[18] Burley JB, Burley CJ. A risk assessment of landscape hazards for building sites in the front range of Colorado. Landscape Research. 1996;**21**(2):137-158. DOI: 10.1080/01426399608706482

[19] Feng M, Burley JB, Machemer T, Korkmaz A, Villanueva MR. Earthquake spatial mitigation: Wenchuan China

and Los Banos, Philippines case studies. GSTF Journal of Engineering Technology (JET). 2018;**5**(2):10

[20] Tokyo Metropolitan Park Association. Kiyosumi Gardens. Tokyo, Japan: Tokyo Metropolitan Park Association; undated

[21] Hong T, Yang W, Zheng Y. Evaluation on the function of urban green space emergency shelter based on AHP. Journal of Fujian Forestry Science Technology. 2013;**40**(1):133-137

[22] Hong T, Wu R, Guo M, Zheng Y. Division of the green space shelter based on the weighted Voronoi diagram. Journal of Northwest Forestry. 2013;**28**(3):255-259

[23] Hong T, Wu R, Song S, Guo M, Zheng Y. GIS based analysis on engergency shelters capacity of urban greenspaces in Sanming City. Journal of Southwest Forestry College. 2103;**33**(1):55-59

[24] Magis K. Community resilience: An indicator of social sustainability. Society and Natural Resources. 2010;**23**(5):401-416. DOI: 10.1080/08941920903305674

[25] Sharifi A. A critical review of selected tools for assessing community resilience. Ecological Indicators. 2016;**69**(1):629-647. DOI: 10.1016/j.ecolind.2016.05.023

[26] Baek J, Meroni A, Manzini E. A socio-technical approach to design for community resilience: A framework for analysis and design goal forming. Design Studies. 2015;**40**(6):60-84. DOI: 10.1016/j.destud.2015.06.004

[27] Tobin G. Sustainability and community resilience: The holy grail of hazards planning? Environmental Hazards. 1999;**19**(2):13-25

[28] Waugh WL Jr. Disaster management for the new millennium: U.S. and Canada politics, policymaking, administration and analysis of emergency management. In: Waugh WL Jr, Sylves RT, editors. Disaster Management in the U.S. and Canada: The Politics, Policymaking, Administration, and Analysis of Emergency Management. Vol. 1996. Springfield, Illinois, USA: Charles C. Thomas Publishers; 1996. pp. 344-359. (Chapter XVI)

[29] Mazmanian DA, Sabatier PA. Implementation and Public Policy. Glenview, Illinois, USA: Scott Foresman; 1983. Public policy analysis and management series

[30] Yeung J. Millions of Animals Are Dying from the Australian Fires, and the Environment Will Suffer for Years to Come [Internet]. Atlanta, Georgia, USA: CNN; 2020. Available from: https://www.cnn.com/2020/01/07/australia/australia-fire-wildlife-deaths-intl-hnk-scli/index.html

[31] Manning W. A National Plan. North Billerica, Massachusetts: USA unpublished draft; 1923. p. 427

[32] United States Geological Survey. Frequency of Damaging Earthquake Shaking Around the U.S. [map]. Washington, D.C., USA: United States Geological Survey; 2014

[33] National Interagency Fire Center. Wildfire Activity by County 1994-2013 [Map]. Boise, Idaho, USA: National Interagency Fire Center; 2015

[34] Federal Emergency Management Agency, editor. Tornado Activity in the United States [Map]. FEMA 320. 3rd ed. Washington, D.C., USA: Federal Emergency Management Agency; 2008

[35] National Oceanic and Atmospheric Administration. U.S. Spring Flood Risk [Map]. Washington, D.C., USA: U.S. Department of Commerce, National Oceanic and Atmospheric Administration; 2017. p. 2017

[36] United States Geological Survey. Volcanic Hazards (Based on Activity in the Last 15,000 Years) [Map]. Washington, D.C., USA: United States Geological Survey; Undated

[37] Environmental Protection Agency. EPA Map of Radon Zones [Map]. Washington, D.C., USA: Environmental Protection Agency; 1993

[38] Galloway DL, Jones DR, Ingebritsen SE. Lands Subsidence in the United States. Washington, D. C: U.S. Geological Survey, Circular 1182;

[39] Radbruch-Hall DH, Colton R, Davies WE, Lucchitta I, Skipp BA, Varnes DJ. Landslide Overview Map of the Conterminous United States [Map]. Washington, D.C., USA: United States Geological Survey; 1978

[40] Schwartz RM, Schmidlin TW. Climatology of blizzards in the conterminous United States, 19592000. American Meteorological Society. 2002;**15**:1765-1772. Blizzards by County, 1959-2000

[41] Johnson R, Burley JB. Snowy range ski resort: An illustration of GIS planning principles. Landscape Architectural Review. 1990;**9**(1):15-18

[42] Burley JB. Visual and ecological environmental quality model for transportation planning and design. Transportation Research Record. 1997;**1549**:54-60

[43] Burley JB, Thomsen C, Kenkel N. Development of an agricultural productivity model to reclaim surface mines in Clay County, Minnesota. Environmental Management. 1989;**13**(5): 631-638

[44] Lewis P. Quality corridors in Wisconsin. Landscape Architecture Quarterly. 1964;**52**(2):275-282

[45] Rusydy I, Faustino-Eslava DV, Muksin U, Gallardo-Zafra R, Aguirre JJC, Bantanyan NC, Alam L, Dakey S. Building vulnerability and human loss assessment in different earthquake intensity and time: A case study of the University of the Philippines, Los Banos (UPLB) campus. In: IOP Conferences Series: Earth and Environmental Science. Bristol, United Kingdom: IOP Publishers Vol. 65, No. 1; 2006. p. 1-12. DOI: 10.1088/1755-1315/56/1/012006

[46] Faidell S, Renton A. Death Toll from Typhoon that Lashed Philippines over Christmas Rises to 47 [Internet]. Atlanta, Georgia: CNN; 2019. Available from: https://www.cnn.com/2019/12/29/asia/philippines-typhoon-toll-intl-hnk/index.html [Accessed: 16 January 2020]

Chapter 7

Planning for Sustainable Tourism Based on Green Infrastructure: A Multiscale Methodology for Revitalizing Depopulated Rural Landscapes

Cláudia Oliveira Fernandes and Giulia Olivetti

Abstract

The abandonment and decline of rural regions are conveying huge challenges to their planning and management. The extraordinary natural and cultural values the landscapes of these regions usually display, coupled with the growing need for urban populations to reconnect with nature, have led to an increased demand for these territories for tourism activities. However, the high sensibility of most of these landscapes requires that the promotion of tourism activities be well thought out and regulated. Therefore, it is necessary to devise strategies to implement a sustainable tourism that can boost the economy and secure the population and, simultaneously, protect the values in presence. In this chapter, a methodology is proposed that approaches the landscape planning from a multiscale perspective, based on (1) a macroanalysis grounded on green and blue infrastructures, (2) the delimitation of areas with distinct landscape character, (3) the identification and requalification of strategic microsites and links following a rural acupuncture concept, and (4) the implementation of a local sustainable tourism network that is expected to have a broad impact in the whole region. The applicability of the proposed methodology was tested in a case study in the north of Portugal, specifically in the Olo river basin.

Keywords: landscape planning, landscape management, depopulation, key landscape points, rural acupuncture, connectivity, sustainable tourism network, institutional cooperation

1. Introduction

Across Europe there is a growing trend of depopulation and landscape degradation in remote rural areas, especially in mountain areas [1]. These regions are then in need for special attention from government entities, and it is urgent to define and implement strategies for their planning and management.

Different management models have been debated over the years, with rewilding gaining momentum. Rewilding proposes a landscape management concept for rural depopulated regions in which active management is reduced and nature is

IntechOpen

allowed to manage itself [2]. However, many of these landscapes are sought after and appreciated due to the judicious balance between the marks of human presence such as villages, terraced fields, shelters, footpaths, etc. and the patches of wild forests and other natural habitats. Thus, it is foremost important to ensure that local populations remain in the region and actively manage the green infrastructure and the mosaic as they are primarily responsible for the special atmosphere that these landscapes emanate.

One of the reasons that lead people to abandon these regions is that traditional agriculture, with low productivity and demanding a great work effort, scarcely guarantees a sufficient income for a comfortable modern life [3]. Therefore, it is up to government entities to support measures that allow the implementation and growth of complementary economic activities to obtain income.

In recent years there has been a growing demand for these regions by urbanites who seek relief from everyday stress in connection with nature [4]. These regions, full of tradition, history, and natural values, can offer, if properly prepared, a diverse range of activities and recreational opportunities for these tourists. However, it is important to note that the high sensitivity of these landscapes is incompatible with the construction of infrastructures of great visual impact such as those usually built up to accommodate mass tourism [5–7].

Sustainable tourism, or ecotourism, has therefore been indicated as an adequate strategy for the revitalization of these regions insofar, as it focuses on the sustainable use of environmental resources, respecting the sociocultural authenticity of the host communities and ensuring sustainable economic operations in the long term. In this model the benefits are provided to all stakeholders, and opportunities, such as the possibility for stable income and employment for the host communities, are appropriately distributed [8–11].

The work presented in this chapter assumes that the implementation of sustainable tourism network must be based on the local green and blue infrastructure, as these already offer a connected network of values, requiring only the identification of the elements (and the links between them), in need of intervention. These interventions are expected to be small scale and low investment, taking place at key components of the metabolism of the landscape. In this way, surgical actions in these components will have a significant and comprehensive impact, far beyond the place where it was operated, influencing the economic and social dynamics of the entire region. That is why we call this assumption—rural acupuncture.

The objectives of this chapter are to propose a methodology for the identification of these landscape components, following a multiscale approach to the landscape, and point out guidelines for their recovery. In the end it should be possible to design a sustainable tourism network, integrating all the cultural and natural values.

The applicability of the proposed methodology was tested in a rural region in northern Portugal, the Olo river basin.

1.1 Sustainable tourism: definition, development, and strategies

Sustainable tourism is a viable sustainable development strategy based on the relationship between man and nature. At the heart of this relationship, there is the correlation between wild and protected areas, with a strong human presence. Specifically, ecotourism can be defined as "Responsible travel to natural areas that conserves the environment and sustains the well-being of local people" [12] and is the fastest-growing sector in the global tourism industry [13]. This growth is primarily due to recognition of the many benefits that this type of tourism can bring to the environment, especially with the renewed concern for climate change and the exploitation of natural resources.

Still, the benefits for local communities, and to tourists actively involved, are gaining relevance [14]. For the environment, sustainable tourism allows a targeted control and protection of natural resources and, at the same time, provides an economic incentive for the implementation of land conservation plans. For local communities, there will be benefits in all territorial levels. For the residents, on a local scale, the advantages will be the development of local trade and the creation of new job positions, encouraging residents not to move to large cities. In addition, investments will be made to protect the environment, which will contribute to the maintenance of biodiversity. At the regional level, local products will enter the market on a wider scale. At the national level, exchange with foreign markets will be crucial, and the capital invested will be retained instead of exported [15].

For tourists, there will certainly be social and psychological benefits. The reconnection with nature, education, and personal growth will lead to greater sensitivity for visitors. Volunteering and financial contribution will also lead to satisfactory activities and may lead to social interactions, which will enable new knowledge and a sustainable lifestyle to be acquired.

However, consideration must be given to the possibility that even sustainable tourism can also adversely affect the environment. First of all, services have to be well managed in order to limit carbon emissions. Moreover, it is possible that local communities, instead of improving and growing economically, are denaturalized; thus it is rather important to preserve their local traditions. Also, profits may not always be used for land conservation funds [14]. For all of these reasons, it is advisable to set up a committee and a plan to regulate sustainable tourism and prevent it from degenerating and damaging instead of protecting.

1.2 Green and blue infrastructure: definition and application in rural land management

Green infrastructure can be defined as a natural or seminatural network of green (soil covered or vegetated like forests, riparian galleries) and blue (water covered by rivers, streams, lakes, oceans) spaces and corridors that maintain and enhance ecosystem services [16, 17]. Its ultimate goal is to support and maintain a healthy ecosystem and the physical and psychological well-being of humans.

This strategy is based on the identification of areas providing ecosystem services, with reference to multiple benefits people can derive from nature [18], creating a network that accumulates and distributes these services throughout the territory [19]. For the implementation of the green infrastructure to be successful, there needs to be collaboration between local communities, landowners, and organizations, to identify, design, and maintain a system, locally, but not only [20].

The importance of green infrastructure is its ability to act on different scales, ranging from local projects to transnational ecological networks. This difference in scale is also reflected in the approach of this strategy, which is based on principles applicable to each type of territory. The basic principles of green infrastructure planning are divided into approaches based on green structures and approaches based on government policies [19].

As for the former, first of all, the integration is important with all the structures and infrastructures present in the territory (e.g., gray structure); the second essential element is the multifunctionality of the strategy, which has an ecological, social, economic, and cultural function.

As far as government policies are concerned, the approach must be long term, with benefits for every single actor involved (local community, landowners, organizations, and government). In addition, social inclusion must be envisaged, enabling the community to participate actively in the various design processes. This will allow a collaboration that will lead to an integrated solution, based on a multidisciplinary approach, allowing it to bring benefits throughout the network.

These benefits may include pollution mitigation, boost in economic value, improvement of built structures, and increase in the scenic value of the landscape.

2. A multiscale methodology to implement a sustainable tourism network

The methodology proposed in this chapter is based on four steps, where the landscape is approached from a multiscale perspective starting from a broader overview; the focus is narrowed down to the identification of strategic microsites, in which surgical interventions will help to revitalize the territory in its entirety. The methodology is graphically represented in **Figure 1**.

2.1 Macroanalysis grounded on green and blue infrastructures

In a first step, a macro and thorough analysis should be conducted in order to gain an in-depth knowledge of the territory, allowing the understanding of all the landscape features.

Biophysical variables, such as geology, slopes, altitudes, the network of natural and man-made waterways, and flora and fauna biodiversity, and socioeconomic variables, such as demography, settlements, and the road network, should be comprehensively analyzed and correlated.

Other features can also help to identify distinct characteristics of the territory under analysis, for example, the prevalence of some tree and shrub species over others and the location and shape of production areas, both natural and man-made, throughout the territory. The state of conservation of the natural and cultural heritage also plays a key role in this landscape characterization.

Moreover, grounded on green infrastructure as a strategy for the recovery and protection of the natural and cultural heritage, it is possible to draw up guidelines for the application of specific measures that help transform tourism into ecotourism. So, first it is important to recognize if there are areas already included in this network and to check if legislation and specific management are required. Furthermore, this structure helps to understand which green corridors exist and which are the waterways with the greatest impact in the territory.

2.2 Delimitation of areas with distinct landscape character

By overlapping the maps produced in previous step, it is possible to identify and classify areas where singular features produce unique landscape characteristics, that is, areas with distinct *landscape character*. These areas as proposed by [21] "display recognisable and consistent pattern of elements that makes one landscape different from another, rather than better or worse." All of the identified areas, from now on called *landscape units*, have to be assessed for their qualities and fragilities; this should allow to understand which approach is best to follow for their revitalization.

For instance, a more "humanized" character will lead to major attention to the repair and valorization of man-made structures, while areas with the prevalence of natural character will shift the focus on the preservation of the native habitats and biodiversity. Usually, in depopulated remote rural areas, a mixed approach is required, demanding attention for both the cultural and the natural heritage.

An important output of this step is, therefore, the deeper knowledge it allows regarding the *landmarks* (natural or cultural components of the landscape, with

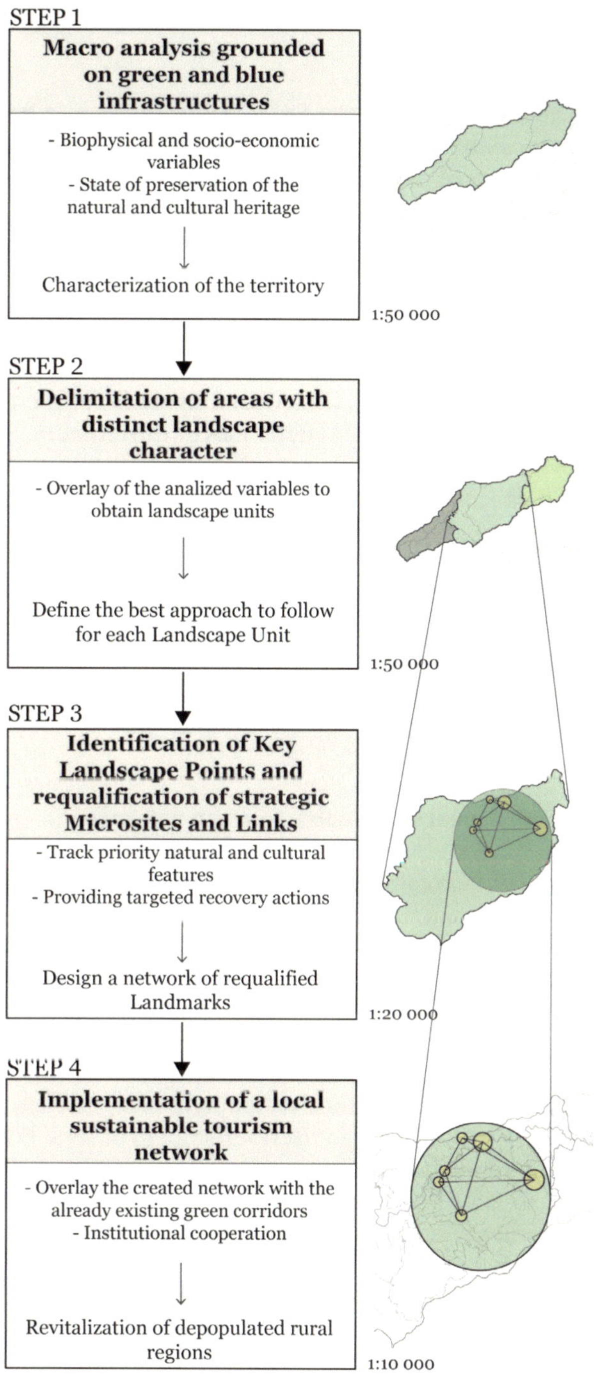

Figure 1.
Schematic diagram of the methodology. Note: The image scales are merely indicative.

high scenic value) of the region, as they will be instrumental for the next step of the methodology.

The definition and mapping of distinct landscape units will also help to identify priority areas for intervention, which, in an ever-present context of scarcity of financial resources for the implementation of these projects, can be important. Another advantage is that by reducing the scale of intervention, it will be easier to plan and implement the needed requalification actions.

2.3 Identification of key landscape points (KLP) and requalification of strategic microsites and links

After having located and characterized each landscape unit, it is now possible to identify *key landscape points*. These are places with a high concentration of landmarks, or where landmarks of extreme importance and rarity are positioned. Also important is to identify and evaluate the state of preservation of these landmarks and the preexisting connections (links) between them and the green infrastructure. These landmarks can be both man-made structures (e.g., bridges, mills, religious structures) and natural components, such as native forests, rocky outcrops, and waterfalls. Furthermore, the road network, as well as trekking and hiking trails, must also be taken into consideration. We call the landmarks in need for any type of requalification actions as *microsites*. Understanding which of these microsites are instrumental for the metabolism and dynamics of the territory, and providing them targeted recovery actions, is the basis of the implementation of a sustainable tourism network. We can, at this point, establish a metaphor with acupuncture, as by intervening in a very small-scale site, we expect to have profound and holistic impact on the whole region.

Considering the high visual sensitivity these landscapes usually display, interventions must be discreet and carefully thought out, in order to preserve the identity and character, both of the landscape unit and of the region. Examples can be taken from recovery of abandoned residential buildings following traditional technics and materials, cleaning of trekking paths, and introduction of information boards.

2.4 Implementation of a local sustainable tourism network

The requalification of microsites and links leads to the creation of a structured grid from which institutions devoted to regional planning and development can cooperate to plan a sustainable tourism network. However, as abovementioned, the implementation of this strategy has also, and primarily, to be based on the green and blue infrastructure (existing green corridors, waterways, forests).

It is expected that the addition of new important tourism services and facilities, further to the improvement and creation of a usable and recognizable pedestrian network, will increase visitation and the development of tourist activities. New activities will bring new jobs, leading to the flourishment of local businesses. This will allow the resident population to continue living in traditional villages, countering the rural exodus. Finally, a sustainable tourism network can also bring benefits to the natural environment. Responsible tourism will not damage the local habitats, flora, and fauna, but, instead, it will create opportunities to increase the knowledge of these values and, thus, the willingness to protect them.

In the next chapter, we will exemplify how the described methodology can be applied in practice.

3. Application of the proposed methodology to Olo river basin

The Olo river basin is located in the north region of Portugal. It covers an area of 13,500 hectares, including the municipalities of Amarante, Mondim de Basto, and Vila Real. The Olo River has approximately 36 km, from its source in Lamas de Olo, Vila Real, to its mouth in the Tâmega River, in Amarante (**Figure 2**).

Approximately 85% of the Olo river basin is part of the Natura 2000 network, namely, the "Alvão-Marão" Community Importance Site (PTCON0003). Moreover, the East section is within a protected area, in the Alvão Natural Park, corresponding to 42% of the river basin.

Most of the population is settled in small remote villages, which have a built heritage that still retains part of its authenticity. However, with the growing depopulation of the inner area of the Olo river basin, most of the villages already exhibit several abandoned buildings, reflecting a progressive degradation of the inhabited centers. Still, even with the progressive abandonment of these settlements, the agroforestry mosaics continue to persist, with traditional systems such as meadows, agricultural terraces, and highland pastures.

Understanding the dynamics of a complex landscape such as the Olo river basin calls for an integrated methodology that encompasses the study of the factors that intervened in its genesis, as well as an analysis of the factors that are responsible for its current situation and its further development. Starting from this, the previously described methodology was then applied to this region.

3.1 Macroanalysis grounded on green and blue infrastructure

This analysis revealed a diverse territory, with a mostly wild and mountainous character, marked by steep topography and relevant waterlines (**Figure 3**).

The hydrography of the Olo river basin is characterized by its high branching, with the most important tributaries being the Ribeira da Fervença and the Rio Sião. These waterlines are vital due to the significant flow, extension, ecological, environmental, and landscape value and also to its importance in the supply of water for villages like Ermelo and Fervença.

Different geological formations are observed. In the central area, the schist predominates, which also abounds in the west, alternating with phyllites. The East Area is dominated by granite, and it is also the area with the highest altitude (>1200 m), sparser vegetation, and where water courses become less frequent.

The valley of the Olo River is then characterized by a rugged relief, dramatic geographical features, and waterfalls of a high landscape, geological, and scientific interest. However, the orography is clearly diluted from East to West, where most of the agricultural activity occurs, predominantly on terraces supported by schist walls.

In the forest cover, mature and expressive riparian forests stand out dominated by *Quercus robur*, *Fraxinus angustifolia*, *Betula alba*, *Salix atrocinerea*, *Corylus avellana*, and *Laurus nobilis*, among other species. Native hardwood and softwood

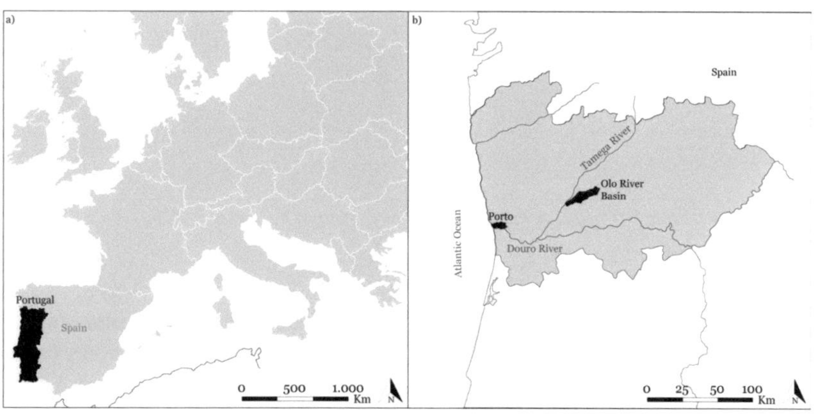

Figure 2.
Left, location of Portugal in Europe; right, location of Olo river basin in the north of Portugal.

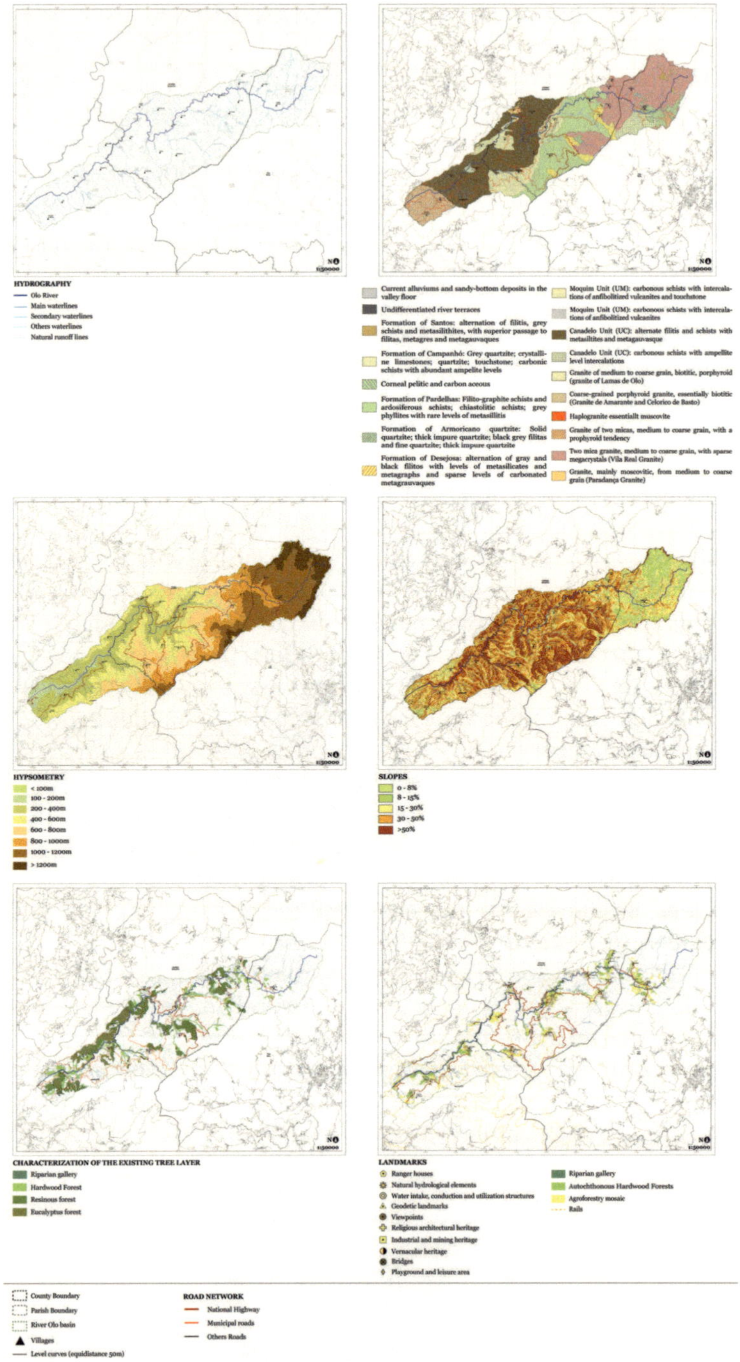

Figure 3.
Map analysis. From bottom left to lowest right: hydrography, geology, hypsometry, slopes, forest cover, and landmarks.

forests dominated by *Quercus* and *Pinus* species are also frequent. Some spots of exotic species (*Eucalyptus globulus*) suitable to produce timber also occur, especially in the northwest area. In higher altitude areas, there is, of course, a dominance of *Quercus pyrenaica* and *Betula alba*. In valley areas with good sun exposure, we also find cork oaks and arbutus trees associated.

Apart from a national road that crosses the majority of the territory, the regional and municipal road networks are poorly maintained and badly distributed, leaving many villages insufficiently served and almost unreachable.

The presence of man is conspicuous but, at the same time, well harmonized with the landscape. Natural and built heritage is abundant, translating the strong relationship between man and nature. The center of this relationship is the Olo River, a forceful natural element that strongly influenced the way man interacted with the territory. Unfortunately, the current degradation of many of these landmarks unveils that this symbiotic relationship is failing.

The built heritage is of "high value," marked by the use of local materials (granite and schist) in the constructions. The houses, granaries, bridges, and the religious heritage are some of the high-value built elements that make up these villages. In the middle of Parque do Alvão, the village of Lamas de Olo, for example, still retains various houses with thatched roofs.

3.2 Delimitation of areas with distinct landscape character: landscape units

The previous analysis exposed evident variations in the landscape character throughout the hydrographic basin of the Olo River. This led to the demarcation of three clearly distinct landscape units, as we can see in **Figure 4**.

These three regions will henceforth be called West Unit, located in the southwest of the river basin; Central Unit, located in the center; and East Unit, located in

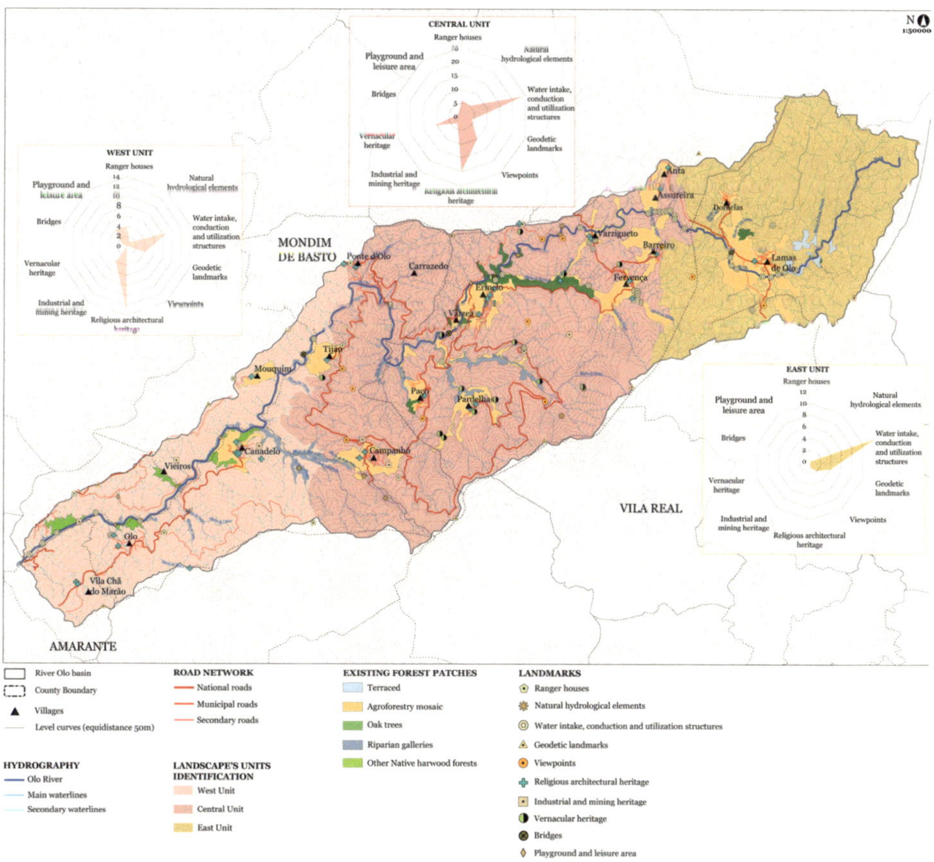

Figure 4.
Synthesis map displaying the landscape units identified in Olo river basin.

the far northeast. The West Unit is mainly characterized by the dominance of flat zones; only along the river, the terrain becomes steeper but always with relatively gentle slopes. We are at low altitudes, an ideal condition for agriculture and timber production. Along the Olo River, riparian galleries with native tree species dominate. It is an area where human settlements have been easier and have had more regular and continuous development over time.

The Central Unit is an area full of slopes and peaks, dominated by watercourses and by an agroforestry mosaic, developed in the surrounding villages. Human intervention is clearly visible in the widespread use of terraces with cultivated fields. Villages always have strategic locations, on top of a hill or in a valley, and are difficult to reach. Thanks to the presence of these differences in altitude, the Central Unit is rich in panoramic viewpoints, which gives it an added importance in terms of its potential for landscape appreciation. As in the West Unit, there are also abundant hardwood forests and riparian galleries, which give the place a special atmosphere.

The East Area is substantially different from the other two. We are at the highest altitudes and the territory is flat. Apart from the Olo River, there are no other predominant waterlines, and the lush vegetation of the open forests allows the development of a shrubbier vegetation adapted to the altitude. Most of this area is part of the Alvão Natural Park. In **Figure 5** two photographs of each landscape unit are presented.

Figure 5.
Photographs of East landscape unit (a) and (b); Central landscape unit (c) and (d); and West landscape unit (e) and (f) (photo credit: Victor Esteves—OH!Land Studio).

The number and quality of landmarks were chosen to help select one of the landscape units for a downsized scale of analysis, in order to progress with the multiscale methodology. Following this approach, it was found that the Central Unit, with 145 landmarks, revealed the highest number, against 60 of the West Unit and 42 of the East Area. Central Landscape Unit was therefore chosen for a detailed assessment of the landmarks.

This unit, with 6832 ha, includes some of the foremost villages of the Olo river basin like Ermelo and Fervença. Fervença village developed close to one of the most important waterlines in this territory, Ribeira da Fervença, along which is possible to appreciate a remarkable forest of autochthonous species and a well-preserved terraced agricultural field. So, in this work, and to better explain the application of the methodology, this subunit Ribeira da Fervença was chosen to be studied in more detail in the next step.

3.3 Identification of key landscape points and requalification of strategic microsites and links

Zooming in Ribeira da Fervença subunit, it was possible to identify six key landscape points that represent places with a high concentration of landmarks needing requalification actions, that is, microsites (**Figure 6**). These landmarks

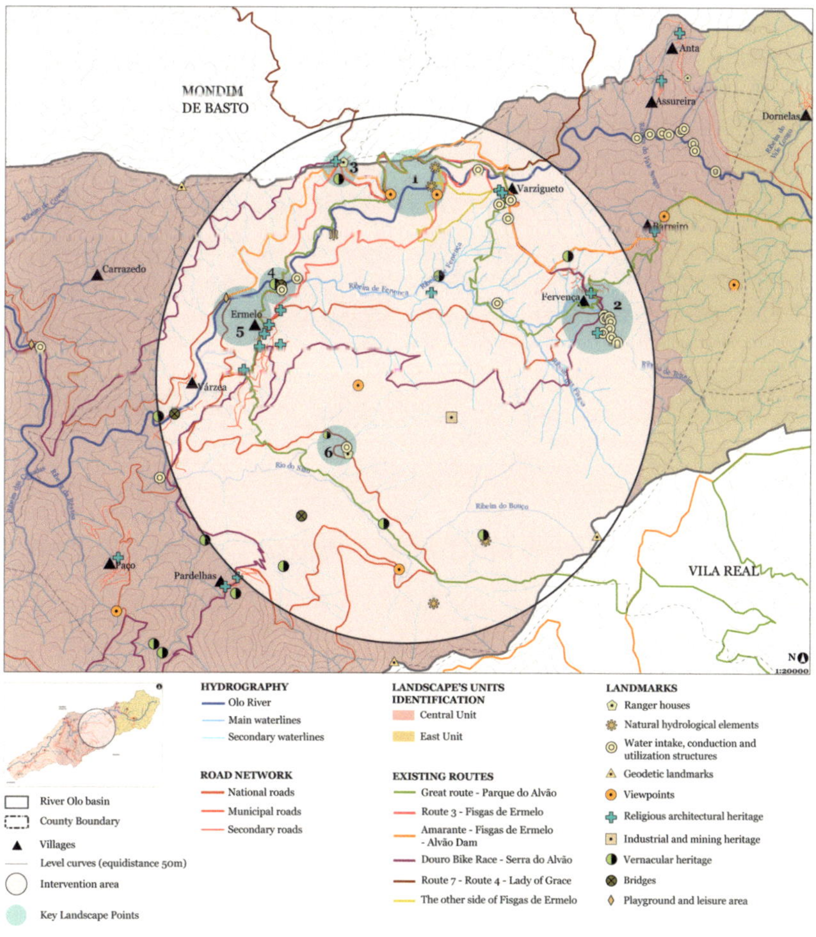

Figure 6.
Location of the six key landscape points in the Ribeira da Fervença subunit.

are prominent elements of the territory's natural and cultural heritage, very well known by the local community, and located along the existing route network.

Each key landscape point reveals a distinct combination of microsites. KLP 1 is rich in natural hydrological elements and in viewpoints; in KLP 2 it abounds water intake, conductions, and utilization structures as well as religious architectural heritage elements; in KLP 3 and KLP 6, it is located in a Ranger's House; in KLP 4 it's possible to observe vernacular heritage and traditional bridges; and KLP 5 includes recreational and leisure areas as well as the important village of Ermelo.

Summarizing, a total of nine different microsites, both natural and cultural, are foreseen for recovery actions. And for each of them, generic goals and specific

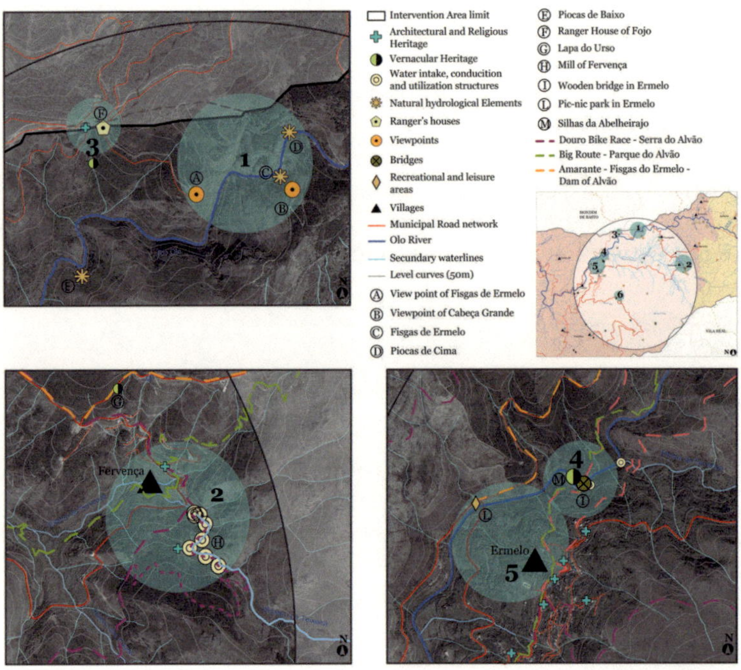

MICROSITES	OBJECTIVES	MEASURES
Rangers' Houses	**Boost regional supply of services and appropriated reception conditions for tourist associated with local values** Figure 8a	1) Requalify and adapt Rangers' Houses to environmental and landscape interpretation centers; 2) Promote recreational and educational activities for tourists through birdwatching, guided tours, plant and animals identification, bike tracking, hiking, etc; 3) Implement training programs to raise awareness on the importance of maintaining agricultural and grazing activities to prevent erosion, forest fires, and biodiversity loss, among other consequences of land abandonment
Recreational and Leisure Areas	**Encourage outdoor activities by improving accesses to better infrastructures** Figure 8b	1) Clean and maintain existing recreation and leisure areas; 2) Implement small green spaces in villages for physical activity, children's leisure, and community use; 3) Conceive new recreation and leisure spots in the vicinity of high environmental quality areas;
Viewpoints	**Promote the sights and scenic value of the landscape** Figure 8c	1) Improve accesses to infrastructures with high environmental and visual quality; 2) Clean and maintain existing viewpoints and introduction of information boards; 3) Install guard rails and requalify existing support equipment
Villages	**Perpetuate local settlements while keeping their historic character** Figure 8d	1) Rehabilitate abandoned residential buildings to accommodate tourists; 2) Renovate built structures of heritage interest, such as threshing floors, granaries, abandoned houses, churches, etc; 3) Improve villages links between them and out of the region
Bridges Vernacular Heritage	**Safeguard the historical and cultural heritage as a factor of identity and tourism potential** Figure 8e	1) Rebuild bridges and other vernacular structures; 2) Include these features in the walking routes plan to be designed
Natural Hydrological Elements	**Preserve the geological and ecological heritage as landscape identity elements, important for tourism promotion** Figure 8f	1) Restore riparian galleries and surrounding areas to be used by the population for low impact recreational use; 2) Develop the pedestrian and road access and increasing security through the installation of secured bodies
Paths	**Upgrade the existing network of walking routes to facilitate local and tourists' access and movement, to and within the region**	1) Design a restoration plan of existing walking routes articulated with other already implemented tourist routes and areas of interest; 2) Expand safe and accessible walking routes, in order to make it possible to visit the most important natural and cultural landmarks of the region;
Water intake, conduction and utilization structures	**Recover of constructed water features related to history and traditions**	1) Repair existing mills, dams, watermills and other water related structures; 2) Modify whenever possible and appropriate, to other functions related to handicraft activities 3) Create new walking routes to pass through cultural heritage

Note: The measures represented in KLP 3 can be also applied in KLP 6

Figure 7.
Location of the microsites in the key landscape points of Ribeira da Fervença and associated objectives and measures for their requalification.

measures to guide the interventions were produced (**Figure 7**). Specific measures can refer to operative improvements, for example, the restoration of the Ranger's House, or to strategic guidelines like those targeting to sensitize the population towards the impacts of field abandonment. Some photomontages were produced to better illustrate the tourist potential of these microsites after requalification (**Figure 8**).

3.4 Implementation of a local sustainable tourism network

Based on the green and blue infrastructure of the Olo river basin, and on the requalified microsites and links, it was possible to design a sustainable tourism network (**Figure 9**). It was confirmed the enormous potential of the Olo river basin for sustainable tourism: It is rich in crystal clear waterlines, and it has impressive waterfalls, a distinct native flora and fauna, very well-preserved habitats, and a long history and rooted traditions. There are plenty of natural blue and green corridors structured on a highly branched hydrographic network and full with riparian galleries.

Figure 8.
Photomontages illustrating the microsites after requalification actions. (a) Ranger's House adapted to landscape interpretation center; (b) refurbishment of a leisure area; (c) viewpoint with security guards and information boards; (d) houses rehabilitated with local materials and technics; (e) rehabilitation of bridges and paths improving walking routes; (f) secure enjoyment near nature and water courses.

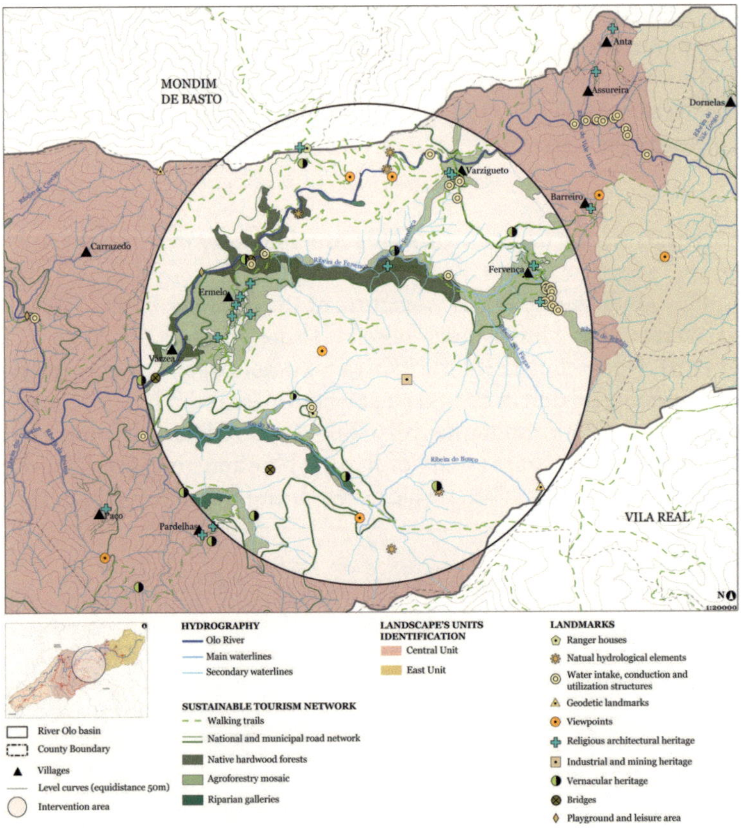

Figure 9.
Implementation of a sustainable tourism network in Ribeira da Fervença subunit.

Furthermore, an agroforestry mosaic very well balanced between terraced green fields supported by small schist walls and forests of native species forests guarantees an ecologic connectivity. To these green and blue infrastructures, an extensive network of unique cultural elements and links, rehabilitated in order to ensure easy and comfortable visitation and recreation, was now assembled. A sustainable tourism network was therefore implemented that simultaneously protects the natural heritage, recovers the cultural heritage, and respects local community's values and traditions.

4. Conclusions

Sustainable tourism is a new way of exploring destinations around the world, without damaging the balance of the planet's habitats and traditional ways of life of local populations. In this chapter it is also argued that sustainable tourism can revitalize depopulated rural landscapes and counter rural exodus by boosting local economy.

A multiscale methodology is here proposed as a strategy to acquire a deep knowledge on the natural and cultural values of rural regions towards the design of a sustainable tourism network. Green and blue infrastructures are assumed as the structuring matrix to which a network of rehabilitated cultural values and improved connections between them is added.

Resorting to the acupuncture metaphor, it is advocated that small-scale, short budget interventions in microsites located in key landscape points will allow a holistic revitalization of the whole region.

The methodology was tested in the Olo river basin, and its applicability was proved, although, for operational reasons, only one of the landscape units has been analyzed. The procedure will be easy to replicate for all identified landscape units, according to an intervention priority. It was therefore demonstrated that the multiscale character of the methodology enables its application in a wide range of landscapes and on different scales, allowing operating on both vast territories and on small regions.

The implementation of this strategy can lead to the creation of a long-term solution for rural remote regions, facing depopulation that is an integrated large-scale sustainable tourism network. This will help limit the waste of resources and minimize the negative impact of too invasive tourism.

A sustainable tourism network has the potential to relaunch local economy, improving the living conditions of the residents and allowing the creation of a very important tourist hub in the region. However, it should be emphasized that the success of any planning strategies is only achieved with the collaboration and active participation of different stakeholders: municipalities, public institutions linked to regional planning and landscape management, tourism-related companies, and small local businesses.

Acknowledgements

The authors would like to thank OH!Land Studio and to the Municipality of Amarante for inspiring and supporting this project. This research was carried out under the project Olo River Living Laboratory for Sustainable Tourism and studies for the Characterization and Enhancement of the Landscape, funded by NORTE 2020 (2014–2020 North Portugal Regional Operational Program) and ERDF 2020 (European Regional Development Fund).

Author details

Cláudia Oliveira Fernandes* and Giulia Olivetti
Department of Geosciences, Environment and Spatial Planning, CIBIO-InBIO—Research Center in Biodiversity and Genetic Resources, School of Sciences, Associate Laboratory, University of Porto, Portugal

*Address all correspondence to: cofernandes@fc.up.pt

References

[1] Viñas CD. Depopulation processes in European rural areas: A case study of Cantabria (Spain). European Countries;**11**(3):341-369. DOI: 10.2478/euco-2019-0021

[2] Three-year Strategic Plan 2019-2020. Rewilding Europe. Amsterdam, The Netherlands; 2019. Available from: https://rewildingeurope.com/wp-content/uploads/2019/05/Rewilding-Europe-Strategic-Plan-2019-2021_website.pdf [Accessed: 25 March 2019]

[3] Fóti K. Quality of life in rural Europe—Key findings from the European Quality of Life Survey. ENRD Workshop on 'Combatting rural depopulation: Creating new opportunities for vibrant rural areas'. 2016

[4] Holden A, Lupton K. Experiencing and connecting to nature: An urban to rural association. In: Slocum SL, Kline C, editors. CAB Linking Urban and Rural Tourism: Strategies in Sustainability. 2017. DOI: 10.1079/9781786390141.0020

[5] Pollock A. Six reasons why mass tourism is unsustainable. The Guardian. 21 August 2013. Available from: https://www.theguardian.com/sustainable-business/six-reasons-mass-tourism-unsustainable. [Accessed: 10 December 2019]

[6] Peeters P, Gössling S, Klijs J, Milano C, Novelli M, Dijkmans C, et al. Overtourism: Impact and Possible Policy Responses. Brussels: European Parliament, Policy Department for Structural and Cohesion Policies; 2018. DOI: 10.2861/919195

[7] Sánchez-Cañizares SM, Castillo-Canalejo AM, Cabeza-Ramírez LJ. Sustainable tourism in sensitive areas: Bibliometric characterisation and content analysis of specialised literature. Sustainability. 2018;**10**:1525. DOI: 10.3390/su10051525

[8] World Tourism Organization. Sustainable Tourism for Development Guidebook—Enhancing Capacities for Sustainable Tourism for Development in Developing Countries. Madrid: UNWTO; 2013. DOI: 10.18111/9789284415496

[9] Anup KC. In: Butowski L, editor. Ecotourism and Its Role in Sustainable Development of Nepal, Tourism—From Empirical Research Towards Practical Application. Croatia: IntechOpen; 2016. DOI: 10.5772/62308 Available from: https://www.intechopen.com/books/tourism-from-empirical-research-towards-practical-application/ecotourism-and-its-role-in-sustainable-development-of-nepal

[10] Dileep MR. Tourism, Transport and Travel Management. Routledge; 2019. p. 470. DOI: 10.4324/9781315151069

[11] Drăgulănescu IV, Druțu M. Rural tourism for local economic development. International Journal of Academic Research in Accounting, Finance and Management Sciences. 2012;**2**(1):196-203. ISSN: 2225-8329

[12] Honey M. Ecotourism and Sustainable Development: Who Owns Paradise? Washington, D.C.: Island Press; 1999. DOI: 10.1177/004728759903800214

[13] Hill JL, Gale T. Ecotourism and environmental sustainability: An introduction. In: Ecotourism and Environmental Sustainability: Principles and Practice. Ashgaten Publishing Limited; 2009. pp. 259. DOI: 10.4324/9781315578767

[14] Ballantyne R, Packer J. International Handbook on Ecotourism. Edward

Elgar Publishing; 2013. DOI: 10.4337/9780857939975

[15] Seba J. Ecotourism and Sustainable Tourism: New Perspectives and Studies. Apple Academic Press, Inc.; 2011. DOI: 10.1201/b12233

[16] Austin G. Green Infrastructure for Landscape Planning: Integrating Human and Natural Systems. Routledge; 2014. DOI: 10.1080/14649357.2014.935608

[17] Naumann S, Davis M, Kaphengst T, Pieterse M, Rayment M. Design, implementation and cost elements of green infrastructure projects. Final report to the European Commission, DG Environment. 2011. Available from: https://op.europa.eu/en/publication-detail/-/publication/42fc9e33-2e83-49f8-a942-412e5a651f0f [Accessed: 06 January 2020]

[18] Millennium Ecosystem Assessment. Ecosystems and Human Well-being: Synthesis. Washington, DC: Island Press; 2005. Available from: https://www.millenniumassessment.org/documents/document.353.aspx.pdf [Accessed: 06 January 2020]

[19] Czechowski D, Hauck T, Hausladen G. Revising Green Infrastructure: Concepts Between Nature and Design. CRC Press, Taylor & Francis Group; 2015. DOI: 10.1201/b17639

[20] Firehock K. Strategic Green Infrastructure Planning: A Multi-Scale Approach. Island Press; 2015. DOI: 10.5822/978-1-61091-693-6

[21] Swanwick C. Landscape Character Assessment: Guidance for England and Scotland. Great Britain. Countryside Agency, Scottish Natural Heritage (Agency); 2002. Available from: https://www.nature.scot/landscape-character-assessment-guidance-england-and-scotland [Accessed: 06 August 2020]

Chapter 8

Methods of Landscape Valorization and Possibilities of Its Application in Hunting Area Categorisation

Szewczyk Grzegorz, Krzysztof Lipka, Piotr Wężyk, Karolina Zięba-Kulawik and Monika Winczek

Abstract

As a result of environmental changes, assessment indexes for the agricultural landscape have been changing dramatically. Being at the interface of human activity and the natural environment, hunting is particularly sensitive to environmental changes, such as increasing deforestation or large-scale farming. The classical categorisation of hunting grounds takes into account the area, forest cover, number of forest complexes, fertility of forest habitats, lack of continuity of areas potentially favourable to wild animals. Landscape assessment methods used in architecture often better reflect the actual breeding and hunting value of a given area, especially in relation to fields and forests. The forest-field mosaic, large spatial fragmentation as well as interweaving of natural environment elements with buildings do not have to be the factors that limit the numbers of small game. Identification of the constituents of architectural-landscape interiors: content and significance assessment, determination of the functional role or assessment based on the general environmental values being represented take into account factors important for the existence of game, in particular small game.

Keywords: landscape valorization, assessment indexes for the agricultural landscape, hunting, categorisation of hunting grounds, deforestation, large-scale farming

1. Introduction

"When you reap the harvest of your land,
you shall not be so thorough that you reap
the field to its very edge,
nor shall you glean the stray ears of grain.
Likewise, you shall not pick your vineyard bare,
nor gather up the grapes that have fallen.
These things you shall leave for the poor and the alien"

(Lev. 19: 9, 10)

Dominant over other species, man has subdued the earth's resources. His expansive economy and, as we know today, often wasteful exploitation of natural resources has been proceeding with varying intensity for several thousand years, practically since the transition from the hunter-gatherer economy to agriculture. Problems of destruction of the natural environment have long been noticed. In Poland, common yew was the first tree legally protected by King Władysław II Jagiełło's decree issued in 1423, which stated: "If a man enters the forest and cuts any trees that are of great value, such as common yew or the like, he may be captured by the lord or squire (...)". The oldest Polish legal regulation regarding the protection of animals was a species protection act issued in the 11th century by King Bolesław the Brave, which prohibited beaver hunting. An office of lord of beavers (*dominus castorum*) was specially created, with beaver guardians (*venatores castorum*) subject to him. They took care of beaver lodges on behalf of the king [1]. Therefore, the protection of natural habitats or similar ones is not characteristic of our times only although it must be admitted that only now, in the era of instant and global information, is it gaining proper significance.

Hunting, which has always operated at the interface between human activity and the natural environment, is particularly sensitive to changes increasingly occurring in the latter: deforestation or the development of large-scale farms.

In Poland, habitat protection is probably most fully implemented in forest areas. This is probably due to the fact that the vast majority of our forests (almost 9.1 million ha in total) are managed by the State Forests National Forest Holding, which manages almost 84% of the country's forest area. Within one entity, it is easier to have a consistent legal framework and uniformity of activities. In accordance with the Forest Act [2], Art. 7.1., the leading goals of forest management are defined by foresters as the conducting of "permanently sustainable forest management (...) taking into account in particular the following objectives: forest conservation, forest protection including the preservation of natural diversity, (...) landscape values (...)". Apparently, the protection should most often apply to wooded areas, but in practice the agricultural landscape brings with it more problems, especially on sites that have not yet been significantly transformed.

Classical landscape valorization methods applied in architecture could be useful in the practice of evaluation of field hunting grounds. "Architecture is the art and the ability to shape and organize space in real forms aimed at satisfying the material and spiritual needs of man" [3]. What functions in space are natural environment systems (ecosystems) and cultural environment ones (human life systems), and landscape is their expression [4]. "As an external expression of the environment constituting a system in space, the landscape will therefore be the most widely understood object of architecture" [4]. The landscape constantly changes under the influence of natural (biotic and abiotic) and anthropogenic factors. There is a vast range of landscape measure systems and many ways of classifying these measures, essentially covering the features of landscape elements: surface area and proportions of classes on the map, number of classes, landscape diversity, shape variation, central zones, isolation, boundary and contrast, landscape fragmentation and analysis of connectivity between landscape elements [5]. Some of these features are already being applied, while others could potentially be used in assessing the quality of habitats in terms of chase game living there.

2. Materials

The analyzes presented in the article were carried out for the hunting model functioning in Poland. The classification of hunting districts used in Poland includes forest hunting areas (where forest land accounts for at least 40% of the cadastral area) and open field hunting areas (where forest land accounts for less than 40% of the cadastral area) (the Act of 13th October 1995: The Hunting Law) [6]. The latter are the overwhelming majority. For example, in the Małopolska Voivodeship there are about 256 hunting districts (as of 2016) with a total area of 1,473,659 ha, where as much as 66% of the usable area is agricultural land. The problems of habitat protection and proper management of such hunting areas should therefore be one of the main objectives of game management. In view of the constant striving to make field areas "productive", what is gaining particular significance are shrubs, small ponds, permanent and periodic wetlands, natural wildlife shelters in fields, roadside tree groups and small meadows, so important for the agricultural landscape. These elements have hydrological as well as protective and feeding values for birds or small game.

Assessment of game habitats could, in a broader perspective, be carried out in two directions: based on methods of valorization of natural environment factors and on assessment of landscape preservation.

3. Methods of landscape valorization

The valorization based on the general natural environment values according to [7] is based on a point system, which assesses, among others: (1) the area occupied by: forests (1 point for every 100 ha), meadows and peat bogs (1 point for every 150 ha), (2) the landscape value: terrain variety: 1–10 points, area of water reservoirs: 1–10 points, river network density: 1–10 points, tree cover density: 1–10 points. Another method of assessment of natural environment values is the method of valorization of ecological usable land in the agricultural landscape, developed by Ilnicki [8]. It is based on the ecological assessment of landscape elements such as ponds and watercourses (surface area, shape, water quality, hydrogeographic conditions, neighbouring vegetation), tree cover density (the occupied part of the water reservoir perimeter, average tree size) and the type of land adjacent to a watercourse. This method can be used to determine the suitability of an area for agrotourism and hunting as well as the effectiveness of the direction of forest and water reclamation and management of degraded areas.

Landscape can be identified based on its selected features, which also leads to its valorization, i.e. assessment and comparison of the values of landscape elements. There are various methods of landscape identification leading to the determination of homogeneous fragments, or units, and their specific landscape. One of the best known methods, developed by Bogdanowski [9], leads to the designation of architectural landscape units and interiors (pl.: JARK-WAK). An architectural-landscape unit is "an area of uniform or very similar shape expressed in units of shape and units of cover, e.g. flat terrain (unit of shape) covered with a chessboard pattern of units filled with gardens (unit of cover)," and consists of architectural and landscape interiors [10]. Valorization consists in the valuation of architectural-landscape unit elements, their division into those of great landscape importance (or lack thereof), "protective" ones, or those subject to degradation. Another method, an impression curve proposed by Wejchert [11], is based on subjective assessment of landscape and urban values on a scale from 1 to 10. Area valorization in terms of ecological values, developed by Chmielewski [12], takes into account, among others, the size of ecosystems and the stability of their functions, biodiversity or scarcity of species occurrence. Another assessment was proposed by Kistowski [13] based on the state of preservation, variety and expressiveness of a given unit, concerning its visual and aesthetic value. Criteria developed by Myga-Piątek and Solon [14] for the purpose of valorization of the cultural landscape in the process of spatial planning include, among others, historicity and uniqueness as well as aesthetic, emotional and functional values. Raszeja [15] uses integrated assessment of the landscape structure, based on landscape indicators proposed by various authors, where the criteria are e.g. complexity, coherence, development level or visual scale [16]. A landscape can also be understood as a mosaic of homogeneous areas (patches), which in the Polish scientific literature have been called "spreads" (pl. *płaty*) by Richling and Solon [17].

For at least a decade, landscape research directions have been described that apply the so-called landscape indexes (metrics) or are based on the concept of ecosystem services. Landscape metrics, calculated on the basis of algorithms implemented in GIS (Geographic Information Systems) software, are based on spatial information in the form of vectorized topographic maps, thematic maps or other criteria in the field of land use and land cover (LULC) mapping. They express, in an objectified manner, various features of the landscape, above all its composition and spatial configuration, as well as allow, based on multi-temporal geodata series, for determination of the dynamics of changes (e.g. the appearance of new patches in the landscape by its fragmentation or total disappearance of its elements) occurring in the analysed landscape [18].

Another approach to the delineation and classification of landscape units can be applied by using a hybrid solution based on an analysis of multi-source and multi-scale spatial data, i.e. vector layers (polygonal, points) and raster layers - using automatic object-oriented image analysis, i.e. the OBIA method (Object Based Image Analysis; [19]). This approach mainly uses satellite imagery (e.g. SENTINEL-2, ESA) or aerial digital photographs derived from both optical sensors, e.g. multispectral (MS) aerial and hyperspectral (HS) imaging, as well as microwave (radar) or so-called LiDAR data (3D point clouds). The latter data provide indispensable, valuable information about the height structure of vegetation, including e.g. occurrence of the shrub layer and saplings as well as stands of complex structure. The application of artificial neural networks for simultaneous segmentation of input images with a negligible role of the operator (who, however, must have extensive substantive knowledge) allows for repeatability and objectivity in the classification of images representing the analysed landscapes with the use of the OBIA (or GEOBIA) approach. The operator controls the segmentation process by determining the rank of the shape and the colour (brightness in individual spectral channels) of the homogeneous pixel groups sought. In addition, the operator sets the maximum size of segments generated by the algorithm, which in the next classification step are combined into appropriate class hierarchies based on e.g. standard deviation of height or NDVI and other variables.

One of the variables that can be used to segment landscape units are the so-called geomorphometric indexes (primary and secondary), generated in GIS software based on precision Digital Terrain Model (DTM). These are available for the entire area of Poland with very high accuracy reaching 10-15 cm (RMSE elevation) both in the form of developed rasters and 3D point clouds (ALS LiDAR) obtainable from the ISOK and CAPAP projects [20]. Dynamic landscape changes are mainly the result of linear investments (e.g. roads, railways) as well as processes related to socio-economic changes occurring in areas mainly used for agriculture. On the one hand, there is a sharp increase in large-scale farms; on the other, what can be observed in areas with poor soil quality is the abandonment of their use and the phenomenon of so-called secondary succession of forest communities [21]. This phenomenon is extremely interesting from the point of view of increasing (in the areas of so-called "agrarian deserts") or decreasing biodiversity (e.g. overgrowing of unused pastures, such as forest glades and mountain meadows, which leads to the disappearance of some plant species and accompanying insects and birds).

The recent decade has been characterised by a dynamic increase in the number of sources and the scope of spatial information regarding the area of Poland, and available especially in the digital form, which can be applied using GIS software [13]. In the context of valorization, the available data sources are in the analogue and raster forms. Particularly helpful are archival and current aerial orthophotomaps, high resolution satellite (HRS) images as well as the Airborne Laser Scanning (ALS) point clouds, which require processing for the purpose of inference and landscape assessment. Laser scanning (LiDAR) is a revolutionary and innovative technology in various fields of science and economy related to monitoring, management and visualisation of the natural environment [20]. Currently, the entire surface of Poland is covered with ALS point clouds, obtained in Standard I (4 points/m^2) for most of Poland or in Standard II, which includes cities (12 points/m^2). What is often applied in current landscape valorization are GIS visibility analyses [22], which allow for the simulation of a view from a selected place based on the Digital Terrain Model (DTM) or Digital Surface Model (DSM). Digital height models enable identification of the variability of field forms along with forests and trees growing there. The use of GIS visibility analyses conducted on 3D data in the assessment of landscape interiors is extremely valuable. 3D models of vegetation and land relief can also be used in the analysis of observation fields (hunting blind platforms) as well as the safety of shooting from hunting weapons (**Figures 1–3**).

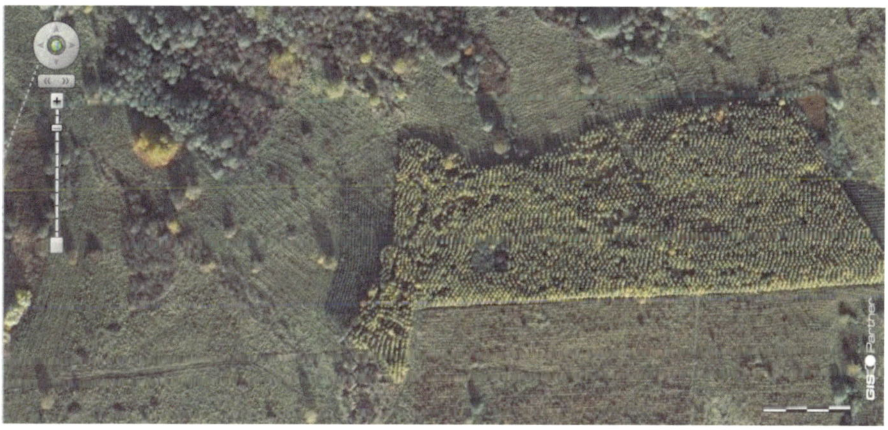

Figure 1.
An RGB aerial orthophotomap.

Figure 2.
A digital surface model (DSM).

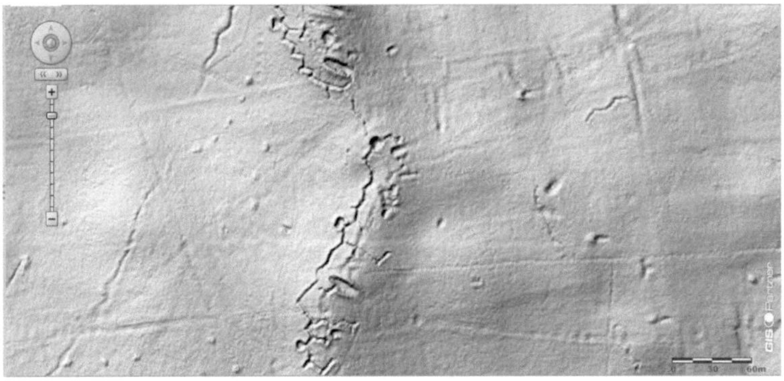

Figure 3.
A digital terrain model (DTM).

Another landscape classification method developed in recent years is based on the concept of ecosystem services (ES) or so-called landscape benefits. Its main categories [23] include:

- supply, e.g. food production, water supply, production of organic raw materials;
- regulatory services (water and air purification, decomposition and detoxification, climate regulation);
- supportive services (nutrient circulation, primary production);
- cultural benefits (spiritual, aesthetic and recreational benefits, scientific discoveries).

4. Hunting in the context of landscape valorization

According to the above classifications, the key problem in Central Europe is a decrease of areas covered by trees and resources of surface, flowing and standing water. For example, the current assessments place Poland among the countries in which the water deficit is going to get worse. It is especially bad for ponds and wetlands. Land drainage works carried out in Poland on a massive scale, especially in the 1960s, have significantly reduced the level of groundwater. The recent dry years and the lack of any work aimed at water retention (deepening overgrown ponds, repairing weirs) have aggravated the long-unfavourable tendencies. In Sweden, eco melioration solutions were introduced in design practice as early as in the years 1980–1990. Straight-line courses of drainage ditches were abandoned in favour of ones which, by meandering, slow down the outflow of water. Adaptation of straight-line ditches to their new functions consisted in creating, in each ditch crown, at 50-meter intervals, 2.5–3-meter-wide enclaves covered with rush vegetation, bushes and low trees. Similar solutions were also proposed for water outlets from drainage system collectors. Such solutions not only improve water conditions in agricultural areas but also significantly valorize the landscape. The most important functions of wetlands still visible in a given area are water retention, water supply to adjacent areas, maintenance of high quality habitats (plant and animal communities), aestheticisation of the agricultural landscape, education and recreation. Despite the currently unfavourable situation, the water retention capabilities of agricultural areas in Poland are still considerable. In an average hunting district with an area of about 4 thousand ha located in an agricultural area in southern Poland, there are about 12 wetlands with a total area of over 27,000 m^2. They retain approx. 40 thousand m^3 of water (**Figure 4**).

In 2006, the vast majority of hunting grounds located in southern Poland still had moderate natural values (valorization class III: 6.5 site index points). Currently, the number of site index points has dropped to 4 (valorization class IV), which indicates low values of the natural environment. By limiting the range of the tree cover and removing the oldest trees, mid-field stretches of land covered by trees and such areas extending along watercourses are degraded to a large extent and fall into valorization class II. This condition is recorded within field hunting areas throughout the region. Such a clear decrease in area quality should be worrying (**Figures 5** and **6**).

If the above division of ecosystem services (landscapes) was used in the aspect of game management in larger areas, individual homogeneous areas should be assigned a specific function or even many functions. Only such a matrix would allow for hunting district valorization in terms of the selected species or a given activity profile.

Many contemporary research issues focus on modelling animal migration routes and species mobility intensity, the occurrence of natural or artificial barriers on a migration route, and the so-called landscape permeability. Identification of land cover patches or plant communities or ecosystems as components of routes (corridors) is helpful in spatial planning at various levels of detail [24], including the construction of animal crossings ("footbridges") over highways and expressways, or as a consequence of rational hunting economy. From the point of view of hunting area categorisation, guidelines for landscape valorization and formulation of recommendations and conclusions regarding landscaping and landscape protection can help in assessing the breeding and hunting value of a given area. Assessed are natural, cultural, historical and architectural, urban, rural complexes,

Figure 4.
Water enclaves with rush vegetation, bushes and low trees.

Figure 5.
Destroyed roadside trees.

Figure 6.
Intensive large-scale agriculture.

including built-up areas distinguished by their local architectural form, as well as aesthetic-visual values, in particular elements of terrain exposure, such as the exposure foreground, view axes and viewpoints. In relation to field and forest areas, cultural features are of some importance; however, the most important are the natural environment features and indicators, e.g. compositional features of plant clusters important for maintaining the diversity of field habitats:

- double or triple clusters - are a transitional form between a lonely tree and a larger cluster of trees;

- gates, frames, wings - these forms are created by trees growing at such a distance that only the edges of their crowns touch. Such a cluster forms view windows which direct the viewing axes; such forms connect landscape interiors;

- clumps - form compact clusters consisting of many trees;
- avenues - are arranger in a linear way in so-called bands forming the walls of a landscape interior; these forms are strongly geometrized and extend along the transport axes;
- lines - types of clusters formed along a line, dividing landscape interiors;
- streaks - they run freely along watercourses or terrain irregularities.

Recognition of the above components of architectural and landscape interiors creates opportunities for content and meaning assessment and for determination of their functional role [9]. Below are some selected examples of assessment:

- a mature linear form that follows natural terrain intersections is more favourable than that which cuts through uniform terrain or opposes its original shape.

Figure 7.
Natural tree group along watercourse.

Figure 8.
Cultural landscape 1.

A classic example are natural tree groups along watercourses. i.e. streaks. Such conditions are extremely favourable for game (**Figure 7**);

- a system of clumps separated by arable fields and rural buildings creates a cultural landscape that is extremely beneficial for small animals and roe deer.

Figure 9.
Cultural landscape 2.

Figure 10.
Linear tree clusters 1.

Figure 11.
Linear tree clusters 2.

Relatively small, varied acreages constitute their food base, and tree groups create ecological corridors. Large-field crops, which have become frequent in recent years, are a particular threat to such an environment. They radically change the food base to one that is beneficial for wild boars, and definitely too poor for e.g. partridges (**Figures 8** and **9**);

- landscape interiors are divided by linear tree clusters, but - assuming the diversification of crops – they determine the separation of homogeneous fragments in terms of form. This way of farming gives animals the ability to move along shelters and reduces the pressure of predators (**Figures 10** and **11**).

5. Discussion and conclusions

In recent years, dynamically progressing changes in the use of field and forest space have been visible throughout Poland. Areas intensively developed with buildings, road infrastructure, power networks, GSM poles and wind farms are growing. They all affect the landscape, changing it irreversibly and leaving their mark. These are certainly factors adversely affecting the existence of game, especially small animals. Changing the landscape from purely agricultural to one enriched with natural succession areas may contribute to an increase in the population of selected animal species, important for hunting.

A varied cultural field-forest landscape is certainly the most beneficial for maintaining the stability of small game populations. The above short presentation of methods of its valorization can be helpful in assessing the quality of hunting districts. Related to the growth of intensive farming economy, there is a visible process of gradual degradation of the environmental valorization class of a given area and a decrease in the value of landscape interiors that determine the living capabilities of game. The cultural landscape shaped over the centuries has quickly managed to reach, in many places, an environmentally and architecturally degenerated landscape.

The Polish Forest Act (1991) rightly points out the close relationship between classic pro-environmental conservation measures and the preservation of landscape values. Therefore, the link between the above activities is landscape ecology, understood as the identification and quantification of relationships occurring between the spatial structure of the landscape (e.g. the number of forest patches) and processes occurring within and between ecosystems [24]. What is particularly important is quantitative assessment of the impact of spatial heterogeneity of the landscape on such phenomena as distribution and movement of animals, which is crucial for conducting sustainable game management as well as active protection of valuable habitats.

Due to the growing human impact on the landscape, in 2000 the Council of Europe adopted the European Landscape Convention, which Poland ratified on 27th September 2004, thus recognising the landscape as an important part of people's quality of life and a key element of the well-being of society. For the needs of the landscape audit, a special typology of landscapes has been developed [14] as well as a classification based in particular on criteria such as: the nature of the factors dominant in the landscape, land relief and land cover. The inventorying of landscape values consists in an analysis and assessment of the values of individual landscapes, taking into consideration the following resources: abiotic, biotic and anthropogenic, i.e. historical, cultural and aesthetic ones [25]. The goal of landscape audits is identification of landscapes which occur within a given voivodeship (province), determination of their characteristics, conducting their valorization and distinguishing the priority landscapes, i.e. those regarded as the most valuable and requiring special protection [26]. An important task is a detailed assessment and presentation of recommendations and conclusions regarding formation and landscape protection, which can be used in many other studies, e.g. in the categorisation of areas managed by hunting clubs. Conducting a landscape audit requires the use of multi-source data, environmental information and field inspections. The detailed results of an audit are to enable its practical application in spatial planning processes at the communal level, or in local strategic-planning documents, including those related to the valorization of landscapes at a microscale. Characteristics of landscapes are prepared by determination of the following analytical features:

1. natural environmental (protected areas and species protection; valuable natural objects; forest and stand site types; boundaries with sea water; ecological corridors; linear bush and tree covers; area fragmentation; land cover; a single field; an agricultural plot; spatial structure);

2. cultural (archaeological sites, rural systems, rural and suburban building objects, objects related to former borders and relict ownership forms, objects connected with fortifications, mining, metallurgy, power industry, craft and industry, religious building complexes and places of worship, places of martyrdom and commemoration; objects of town and palace architecture; historical objects of architecture connected with transport, spa, tourism and recreation, leisure, sports, observation and navigation infrastructure; protected objects; communication objects);

3. synthetic (tradition, identity, familiarity, basic and supplementary functions of a landscape).

Visible in recent years across Europe, a decline in small game has attracted the attention of practitioners and scientists to the causes of this regression. An indication of the reasons for this phenomenon constitutes the starting point for possible corrective actions; the next task should be to assess (valorise) adverse effects. This assessment is extremely important, because by ranking threats, it sets a schedule for corrective actions. Research methods that could be used in the situation described are different than in the case of standard assessments used at the stage of categorisation of hunting districts: area, forest cover, number of forest complexes, fertility of forest habitats, lack of continuity of a given district. The factors included in it do not necessarily have to reflect the actual value of a field area. In relation to small game, the field-forest mosaic, large spatial fragmentation and interweaving of natural environment elements with buildings do not have to be the factors limiting its number. Hunters, obviously interested in maintaining such game numbers that are appropriate for the capacity of hunting areas, should be particularly sensitive to the landscape values of the area in which they hunt. Contrary to appearances, this factor creates an inseparable whole with environmental components.

Author details

Szewczyk Grzegorz*, Krzysztof Lipka, Piotr Wężyk, Karolina Zięba-Kulawik and Monika Winczek
University of Agriculture in Krakow, Kraków, Poland

*Address all correspondence to: rlszewcz@cyf-kr.edu.pl

References

[1] Kurzępa B. 1999. Ochrona zwierząt. Przepisy, piśmiennictwo. Bielsko-Biała, ss. 163-164.

[2] Ustawa o Lasach. 1991. Dz.U. 1991 Nr 101 poz. 444

[3] Encyklopedia PWN. https:// encyklopedia.pwn.pl/

[4] Łuczyńska-Bruzda M. Elementy Naturalne środowiska. Politechnika Krakowska; 1996

[5] Urbański J. 2012. GIS W BADANIACH PRZYRODNICZYCH, Wydawnictwo Uniwersytetu Gdańskiego, str. 182-207.

[6] Ustawa Prawo Łowieckie. 1995. Dz.U. 2015. 2168

[7] Oświt J, Dembek W. Wstępne zasady waloryzacji przyrodniczej mokradeł i związanych z nimi krajobrazów. [w:] Torfoznawstwo w badaniach naukowych i praktyce. Sesja naukowa z okazji jubileuszu 45 lecia działalności naukowej i 70 rocznicy urodzin prof. Dr hab. Henryka Okruszko. Falenty 6-7 XI 1995. Mater. Semin. Wydawnictwo IMUZ. 1995;**34**:121-127

[8] Lipka K, Zając E, Klatka S. Waloryzacja przyrodnicza zrekultywowanych terenów pogórniczych zakładów przemysłu gipsowego "Dolina Nidy" w Gackach. Zeszyty Problemowe Postępów Nauk Rolniczych. Z. 2007;**519**:189-197

[9] Bogdanowski J. Łuczyńska-Bruzda M., Novák Z. 1973. Architektura Krajobrazu. Kraków: PWN;

[10] Bogdanowski J. Problemy Metodologiczne Rewaloryzacji Urbanistyczno-Krajobrazowej Miasta Zabytkowego na przykładzie Kazimierza Krakowskiego. Warszawa: Wydawnictwo Pracowni Konserwacji Zabytków; 1985

[11] Wejchert K. Elementy kompozycji urbanistycznej. Arkady. 1984:1-278

[12] Chmielewski T. System Planowania Przestrzennego Harmonizujacego Przyrode i Gospodarke. Politechnika Lubelska; 2001 Lublin, t. 1: ss. 294

[13] Kistowski M. 2014. Źrodła danych, W: Przygotowanie opracowania pt. "Identyfikacja i ocena krajobrazów – metodyka oraz główne założenia", PAN IGiPZ, str. 10-30.

[14] Myga-Piątek U., Solon J. 2014. Wyrożnianie krajobrazów w obrębie województwa, W: Przygotowanie opracowania pt. "Identyfikacja i ocena krajobrazów – metodyka oraz główne założenia", PAN IGiPZ, str. 30-42.

[15] Raszeja E. Ochrona Krajobrazu w Procesie przekształceń obszarów Wiejskich, Wydawnictwo Uniwersytetu Przyrodniczego w Poznaniu. Poznań; 2013

[16] Solecka I. Polskie doświadczenia w identyfikacji i waloryzacji krajobrazu. Inżynieria Ekologiczna Tom Nr. 2016;**50**, str:223-231

[17] Richling A. Solon J. 2002. Ekologia Krajobrazu, PWN.

[18] Wężyk P, Tracz W, Guzik M. Evaluation of Landscape Structure Changes in Tatra Mountains (Poland) Based on the 4D GIS Analysis. Remote Sensing and Geographical Information Systems for Environmental Studies. Application in Forestry. Schriften Aus der Forstlichen Fakultät der Universität Göttingen Und der Niedersächsischen Forstlichen Versuchanstalt. Band 138. J.D. Sauerlaender'sVerlag. Frankfurt am Main, ISBN 3-7939-5138-3; 2005. pp. 224-232

[19] de Kok R, Wężyk P. Principles of full autonomy in image interpretation. The basic architectural design for a sequential process with image objects. Object-Based Image Analysis. 2008

[20] Wężyk P. 2014. Przyszłość technologii lotniczego skanowania laserowego. W: Wężyk P. (Ed.) Podręcznik dla uczestników szkoleń z wykorzystania produktów LiDAR. Warszawa, s. 320-328, ISBN: 978-83-254-2090-1.

[21] Wężyk P., de Kok R. 2005. Automatic mapping of the dynamics of forest succession on abandoned parcels in south Poland. Strobl et al. Eds. Angewandte Geoinformatik 2005. WichmanVerlag. Heidelberg; ISBN 3-87907-244-4, 774-779.

[22] Zięba K, Wężyk P. 2016. The landscape recomposition of the FestungKrakau - a new approachbased on Airborne Laser Scanning point cloud processing and GIS spatial analyses CRACOW LANDSCAPE MONOGRAPHS. 2016;3:183-191

[23] Richling A. 2014. Główne kierunki badań nad krajobrazem. Problemy Ekologii Krajobrazu, T. XXXIII, pp. 9-15.

[24] Solon J., Pomianowski W. 2014. Program GraphScape – nowe narzędzie do analizy struktury przestrzennej i stopnia łączności w obrębie krajobrazu. GraphScape software – a new tool for analysing landscape spatial structure and connectivity. PEK, T. XXXVIII.

[25] Chmielewski T., Myga-Piątek U., Solon J., 2014. Charakterystyka wyróżnionych krajobrazów, W: Przygotowanie opracowania pt. "Identyfikacja i ocena krajobrazów – metodyka oraz główne założenia", PAN IGiPZ, str. 43-83.

[26] Solon J. 2014. Sposób i zakres formułowania zaleceń dotyczących zarządzania Krajobrazem, W: Przygotowanie opracowania pt. "Identyfikacja i ocena krajobrazów – metodyka oraz główne założenia", PAN IGiPZ, str. 87-90.

Chapter 9

Assessing Ecosystem Services Delivered by Public Green Spaces in Major European Cities

Rui Alexandre Castanho, José Cabezas, José Manuel Naranjo Gómez, José Martín Gallardo, Luis Fernández-Pozo, Sema Yilmaz Genç, Sérgio Lousada and Luís Loures

Abstract

In the last decades, there was a significant population growth in urban areas. In this regard, the European major cities are not an exception; in fact, they are even still more affected by that populational exodus and consequently for an urban growth. Therefore, and considering that the urban parks in the cities are not growing at the same pace, a question is raised: "Are the public green spaces in the European major cities still able to provide the needed ecosystem services to their populations?" Based on the above-mentioned question, the present chapter aims to provide the first insights and answers to this question. Contextually, the study uses a case study research (CSR) method over several European major cities. Besides, GIS tools crossing statistical data are also used to analyze the data and consequently understood and establish a state of the art regarding this relevant issue.

Keywords: ecosystem services, landscape planning, sustainability, urban green spaces

1. Introduction

The original landscapes of our planet have been undergoing transformations by human activities. In Europe, a large part of the original forests existing during the human hunter-gatherer stage has been replaced by agricultural territories and large cities. At the same time, there is remarkably an uneven distribution of the population that results in very low densities in some territories, rural, and very high in urban areas, where significant percentage of inhabitants has been concentrated, throughout a process that has gone developing in the last 150 years [1]. The city was the focus of growth of the states, due, in large part, to the industrialization that led to an increase in the economy, which in turn led to a very rapid expansion and a first concentration of the industries and then of the services. But this great growth caused a disorganized and chaotic development.

Urban planning techniques try to eliminate and prevent urban chaos. In this context, when comparing the pre- and post-industrial revolution growth of the cities, a key difference appears "(...) compared to the old cities with clear boundaries enclosed by walls, post-industrial revolution growth leads to the invasion of the surrounding landscape [2]." The exterior goes from being a threat to the city to being an element threatened by it. The city has evolved in recent centuries toward the need to develop an urban planning concept in which the existence of green spaces became more important. The Industrial Revolution caused the exodus from the countryside to the city and the emergence of epidemics related to lack of health; together with the growing demand for leisure and free time by the population, the need for public green spaces increased.

The Urban Parks Movement (eighteenth century) appears, whose objective was to recreate the presence of nature in the urban environment, in order to improve the quality of life of its citizens [3, 4]. This concept resulted in the creation of the main parks, the first of them in the United Kingdom: "Victoria Park" in London and "Birckenhead Park" in Liverpool; a little later, also in London, "Hyde Park" and "St. James Park"; while in Paris the "Bois de Boulogne" and "Bois de Vincennes" were built and in Madrid "El Retiro." Urban green spaces are urban areas in which natural or seminatural ecosystems became urban spaces by human influence [4]. They provide a connection between the urban and nature [5]. Green spaces include street trees, green roads, green roof walls, urban parks, and even abandoned unbuilt land. In fact, its creation can be from scratch, modified from existing vegetation, generated by colonization or existing as a natural enclave [6]. Vegetation in cities has multiple benefits that have been the subject of vindication and study throughout the evolution of current urbanism and that have been enriched and concretized by the contribution of research from related fields such as ecology.

The presence of abundant vegetation in cities is ideal with a universal appeal, which goes beyond temporal, spatial, and cultural divisions, associating itself with the concept of environmental quality, which leads to a better quality of life. In recent years there is an important interest in the environmental benefits of green spaces. Thus, a significant number of studies attempt to demonstrate, quantify, and incorporate them into planning. However, they still coexist with the marginality which they are treated in practice [7]. The presence of natural elements and values in the city is today a fundamental condition for the environmental recovery of urban territory. The natural and urban systems are part of the same space, and their integrated management is a requirement of the regional space and a condition of sustainability of the territories and cities. In addition, the agroforestry existence in the peripheries of cities and green spaces within the urban fabric represents an increase in environmental quality, which urban planning must strengthen and improve [3, 8–11].

The visual approach of the green areas constitutes a powerful tool to activate and inspire the daily life of citizens. Besides, a deeper understanding of the ecological processes that occurs in nature, along with the economic and socio-cultural, can help city managers to better integrate all the above-mentioned aspects. This approach must go beyond the superficial, appreciating the stories that landscapes tell and helping to understand the place of humans in nature [12].

Studies on the valuation of ecosystem services (ES) focused on urban areas represent a small percentage in relation to the total number of articles devoted to the subject. Furthermore, Delgado and Marín [13] analyzed the growth of publications in a 24-year interval (from 1990 to 2013), demonstrating their exponential growth, which increased from 1 article in 1991 to less than 250 in 2007 and 1500 in 2013. Of these, only 6% focused on the direct services of the ecosystems associated with urban areas. According to Ibes [14], the valuation of the ES was originally designed for non-urban systems, so that new models are necessary for a correct assimilation of the services provided by the urbanized environments. In addition, it reflects on the difficulty of finding a balance between geographical, conceptual, and spatial considerations when the ES valuation paradigm applies to urban parks. Therefore, bearing in mind that urban parks cannot generate all the possible ES, excluding the necessary compensation, it will lead more often in losses rather than benefits.

The key components that contribute to the total economic value of ES can be divided into three main blocks [15–16]. The first is related to the direct use and includes both (a) the provision of services (e.g., the production of plant and/or animal biomass) and (b) social and cultural services (e.g., recreational activities, sports, family). The benefits associated with urban parks are mainly framed in the second group, presenting the contributions to the first residual character in general. The second block refers to indirect services (indirect use) that involve (c) regulating (such as the control of air, water or soil quality) and (d) supporting services that are necessary for the production of the rest of services of the ecosystem (e.g., nutrient cycles, soil formation, or water cycle).

The parks contribute to a greater extent in the section of regulating services, with benefits that include the improvement of the air quality or the decrease of the load of nutrients that reach the water courses and are potential causes of eutrophication. The third block is dedicated to other aspects not contemplated in the previous ones. It comprises two sections: (e) option services, referring to the possibility of using a service in the future and maintaining resilience (ability to reverse changes in the ecosystem) and (f) nonuse/exploitation of resources of ecosystem resources for cultural reasons and of preservation for future generations or their intrinsic values. The ES of urban parks contribute more to the aspects related to the second section.

Several authors have evaluated the benefits of the parks valuing some specific ES. Also, Breuste et al. [17] analyzed in three megalopolis the importance from the recreational point of view (Buenos Aires and Shanghai) and climate regulation (Karachi). They demonstrated that urban parks play an extremely important role by offering ES related to recreation and contact with nature. With regard to Karachi, they highlighted the importance of parks in the regulation of extreme weather conditions. Residential areas located near parks had a considerable higher degree of thermal comfort. Setälä et al. [18] assessed the retention of heavy metals and nutrients in the soil, highlighting the role of parks especially in cities with high levels of pollution. Regarding the contribution of the ES in urban parks, Gratani et al. [19] studied and quantified four parks located in Rome to carbon sequestration. Mediterranean-type parks, such as the Romans, sequestered CO2 throughout the year highlighting the results in those in which the native species of the Mediterranean basin were dominant. The annual economic value of the CO2 elimination would be equivalent to \$23,537 ha^{-1}. Moreover, Giedych and Maksymiuk [20] studied the Warsaw parks, concluding that the ES contributed by each of them depend on the local conditions and specific characteristics of each of them, the surface being one of the key variables in the regulating services.

Less abundant are the works that analyze and value the set of ES that generate concrete parks. An example would be the holistic valuation of the ES generated by Central Park (New York, USA), estimated at \$ 70 million/hectare/year [21]. Contextually, the present chapter through a case study research method aims to analyze the green urban areas surfaces evolution in seven European major cities.

2. Materials and methods

Initially, land use data monitored by Land Cover Corine (CLC) were obtained (https://land.copernicus.eu/pan-european/corine-land-cover) [22] on a scale of 1:100,000, with a minimum mapping unit (MCU) of 25 Ha and using polygonal graphics features that evoke land uses in Europe. Some of the used CLC nomenclature/codes used are shown in **Table 1**.

In addition, the urban boundaries of the cities analyzed were obtained from ESRI-free data, using a layer called Europe Shapefiles. In this case, polygon features were also used.

In this regard, the authors have analyzed these two layers of information – which represent two variables in the same georeferenced position. For this reason, the two layers were transformed into the same reference system, using ETRS1989 Lambert azimuthal equal area, because this projection preserves the areas and is better suited to the different cities to be analyzed.

From the two polygonal cartographic layers, an intersection was made between the two. Thus, polygons corresponding to land uses that are completely included in the boundaries of cities become part of the resulting layer. Also, the parts of the polygons corresponding to the land uses that are partially included and clipped by the boundaries of the cities are also part of this resulting layer. Thus, it was possible to obtain a layer with the land-use polygons within each city.

Once this layer was obtained, we proceed to measure the surface of each of these polygons obtained evocative of the land uses, but in the projection used. In order to do this the ArcGIS 10.3 software was used. Subsequently, using Microsoft Access 2016, selection queries were made. Thus, only polygons whose use was 1.4.1 corresponding to Green urban areas were chosen, that is to say, areas with vegetation urban fabric which includes parks and cemeteries with vegetation. Later, a query was carried out so that the total area dedicated to green urban areas was obtained.

Therefore, seven case studies of European major cities were selected (**Figure 1**). After the case study selection, an analysis for the years 1990, 2000, 2006, 2012,

Level 1	Level 2	Level 3
1 Artificial surfaces	11 Urban fabric	111 Continuous urban fabric
		112 Discontinuous urban fabric
	12 Industrial, commercial, and transport units	121 Industrial or commercial units
		122 Road and rail networks and associated land
		123 Port areas
		124 Airports
	13 Mine, dump, and construction sites	131 Mineral extraction sites
		132 Dump sites
		133 Construction sites
	14 Artificial, nonagricultural vegetated areas	141 Green urban areas
		142 Sport and leisure facilities

Table 1.
CLC nomenclature [22].

Figure 1.
Selected case studies. (A) Lisbon, (B) Madrid, (C) Paris, (D) London, (E) Rome, (F) Berlin, (G) Stockholm.

and 2018 was carried out. Nevertheless, for the cities of London and Stockholm for 1990, there was no data.

Finally, thematic maps representative of land uses were obtained for each of the years and cities, highlighting the green urban areas.

3. Results and discussion

From the 11 classes of the CLC, the study only analyzes Level 3 (land use code 141)—regarding green urban areas (**Table 1**). Those results were presented in acres and were assessed for each year of the studied period (1990, 2000, 2006, 2008, 2012, and 2018) (**Table 2**). Contextually, the results presented in **Table 2** enabled to create a graph (**Figure 2**). This graph shows the cities being grouped into three levels. In the first level, we have London with the largest surface of green areas over the studied years—around 12,000 acres. On a second level, we have Stockholm, Madrid, and Paris that slightly have a surface of green urban areas superior to 4000 acres; however, any of those reach the 8000 acres. In this regard, it should be highlighted that in the first studied year (1990), Madrid was one of the cities with lowest values regarding green urban areas surfaces, and in the last studied year (2018) the Spanish capital reaches the third position—as one of the studied cities with the highest value of CLC 141. And in a third level, we have the studied cities with the lowest values of green urban areas, which are Berlin, Lisbon, and Roma, with less than 4000 acres of the land use 141—in fact, with a CLC 141 surface lower than 2000 acres.

Moreover, through the creation of individual graphics for each of the selected cities, it was possible to analyze in detail how the green urban area surfaces evolved over the 5 years studied (**Figure 3**). Through this analysis, it is possible to verify that two cities (Rome and Stockholm) are losing green

City	1990	2000	2006	2012	2018	Dif.	%
Berlin	2896.32	2868,46	2873,66	3102,04	3336,18	439,87	15,19
Lisbon	1204,02	1465,01	1827,66	1783,96	1929,92	725,91	60,29
London	n.d.	11,429,73	12,380,38	12,195,22	12,224,16	794,43	6,95
Madrid	2337,87	3246,07	5798,95	6457,62	6428,88	4091,01	174,99
Paris	4564,59	5183,53	5212,09	5239,16	5187,85	623,26	13,65
Rome	1654,96	1532,34	1456,55	1456,55	1455,86	−199,09	−12,03
Stockholm	n.d.	6954,17	6907,24	6901,44	6869,19	−84,98	−1,22

Table 2.
Outcomes of the analyzed parameters of the green urban areas in European major cities (source: Authors). n.d., no data available; dif., difference between first and last year; %, percentage.

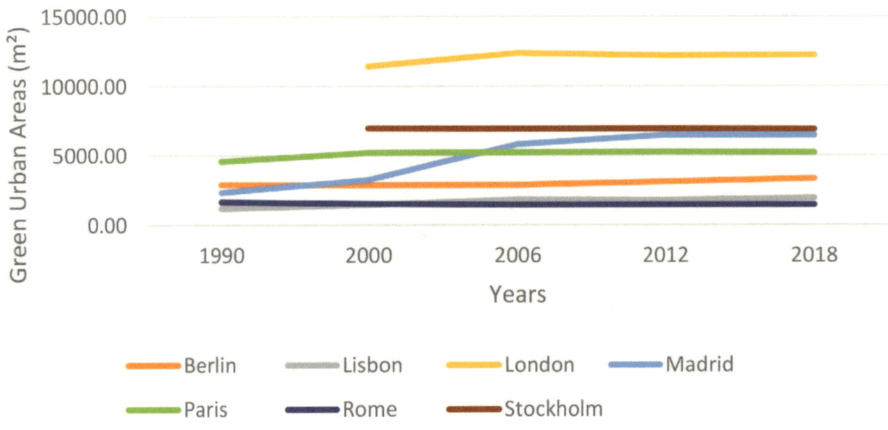

Figure 2.
Evolution of the urban green spaces through the years in the studied European major cities (authors).

urban areas in comparison with the first year analyzed (1990). On the other hand, all the other cities are gaining more green urban areas along the years. From all those cities that show an increase in the land use 141 over the years, it should be highlighted that Madrid and Lisbon show constant growth. In fact, this tendency was also identified in Berlin; however, it only starts in the year 2012 onwards—once the German capital presented a period of growth stagnation (of the land use 141) in the previous years. Besides, in Paris and London, we have been identified the opposite scenario. In Stockholm, the city was lost Green Urban Areas surface when compared to the 1990 reality; however, it was also started a similar growth process (regarding the land uses 141) in the year of 2012 – which is verified in the year of 2018; in an opposite tendency, we have the city of London. The city of London, even it has been passed through an increase of Green urban Areas in the first year studied (1990), is now facing a tendency of decrease in these green surfaces—which started in the year 2006.

Regarding the results in percentage (**Table 2**), Roma and Stockholm have lost 12.03 and less than 1,22% of their green urban area surfaces, respectively. In contrast, the cities that gained more green urban areas have been Madrid, with 174,99%, and followed by Lisbon, Berlin, and Paris (between 60,29, 15,19, and 13,65%). Furthermore, London increases its land use 141 in less than 5%, nevertheless, with a negative tendency (**Table 3**).

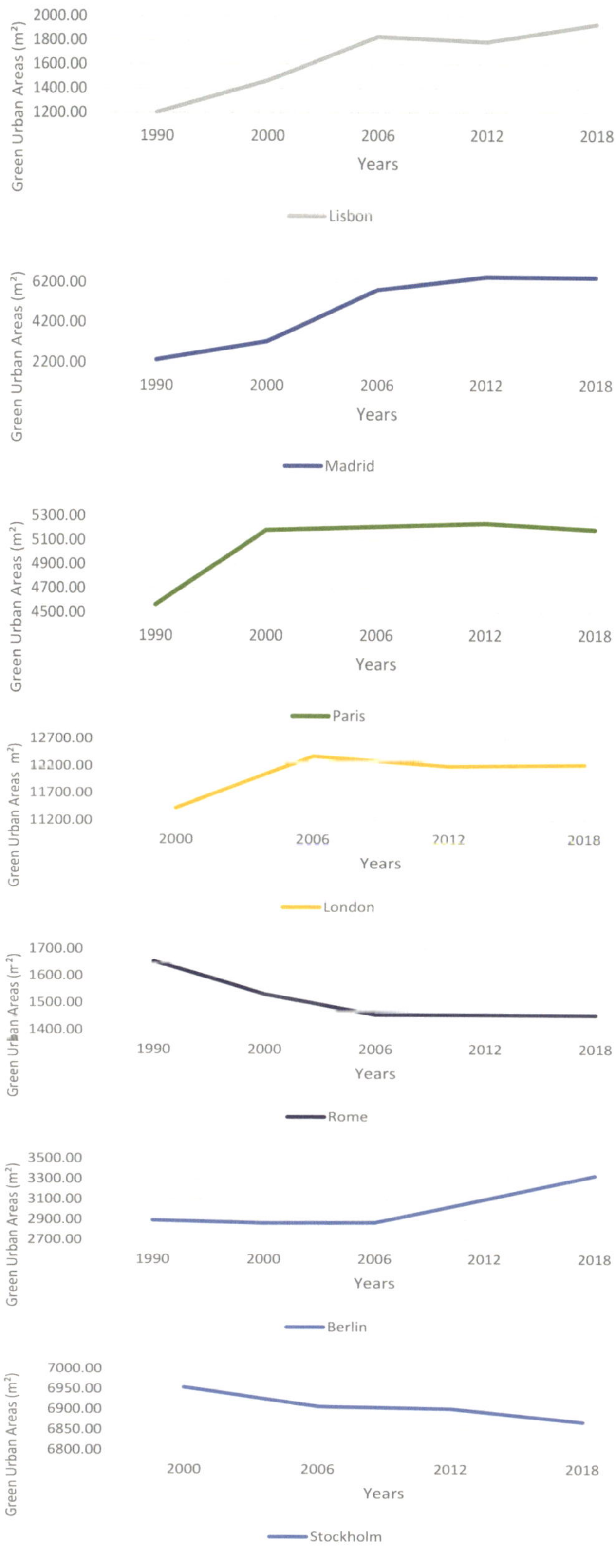

Figure 3.
Evolution of the urban green spaces in European major cities (authors).

Case studies	Population (thousands)			
	1990	**2000**	**2010**	**2015**
Stockholm	1030	1210	1360	2615
Madrid	4940	5320	6380	6586
Rome	3750	3710	3960	4468
Paris	2150	2130	2240	12,524
London	6800	7240	8600	14,855
Berlin	3200	3500	3450	4314
Lisbon	2540	2690	2790	2810

Table 3.
Demographic dynamics of the studied cities [23].

4. Final remarks

Through the present study, it is possible to understand how the green urban areas have evolved within the studied European major cities. Besides, throughout the analysis of patterns of the land use change (CLC 141) along with empirical knowledge of those cities' territories, it was allowed us to assess the value of those Green Urban Areas within the cities. Therefore, it is possible to say that those green urban areas are not growing in the same pace as the demographic values as well as other land uses in development within these cities [24].

In this regard, and considering the relevance of the ES performed in the urban environments, we believe that in all the analyzed cities, the existing green urban areas are not able to provide the environmental needs for their inhabitants. In fact, even if those environmental needs could differ among the studied cities – once, some presents a higher number of Green Urban Areas than others as well as different demographic growth rates; all the analyzed European Major Cities shows a need for more Green Urban Areas.

Additionally, the performed study enabled us to put forward some noteworthy ideas, related to the relevance of green space infrastructure in urban areas, regardless of their urban nature and of their major land use, which corroborate with the conclusions of previous studies that crossed the relevance of urban green spaces to urban sustainability and development [4, 9–10, 25–32].

In this regard, the creation of more green urban areas in these cities as well as in their metropolitan influential territories is seen as pivotal. Furthermore, guidelines should be provided for the main actors and decision-makers of the planning process to where the efforts toward a sustainable development and growth should be placed—for example to address green strategies and land use reconversion and redevelopment of urban areas.

Author details

Rui Alexandre Castanho[1,2,3,4,5]*, José Cabezas[2,4], José Manuel Naranjo Gómez[3,4,6], José Martín Gallardo[7], Luis Fernández-Pozo[2,4], Sema Yilmaz Genç[8], Sérgio Lousada[3,4,9] and Luís Loures[4,10]

1 Faculty of Applied Sciences, WSB University, Dąbrowa Górnicza, Poland

2 Environmental Resources Analysis Research Group (ARAM), University of Extremadura, Badajoz, Spain

3 CITUR - Madeira - Centre for Tourism Research, Development and Innovation, Funchal, Madeira, Portugal

4 VALORIZA - Research Centre for Endogenous Resource Valorization, Polytechnic Institute of Portalegre (IPP), Portalegre, Portugal

5 School of Business and Economics and CEEAplA, University of Azores, Ponta Delgada, Portugal

6 Agricultural School, University of Extremadura, Badajoz, Spain

7 Functional Studies of Mediterranean Ecosystems, University of Extremadura, Badajoz, Spain

8 Department of Marketing and Advertising, Kocaeli University, Turkey

9 Faculty of Exact Sciences and Engineering (FCEE), Department of Civil Engineering and Geology (DECG), University of Madeira (UMa), Funchal, Portugal

10 Research Centre for Tourism, Sustainability and Well-being (CinTurs), University of Algarve, Faro, Portugal

*Address all correspondence to: alexdiazbrown@gmail.com; acastanho@wsb.edu.pl

References

[1] Pinilla V, Ayuda MI, Sáez LA. Rural depopulation and the migration turnaround in Mediterranean Western Europe: A case study of Aragon. Journal of Rural and Community Development. 2008;**3**(1):1-22

[2] Aponte G. An approach to landscape planning in Borders. In: Landscape Planning Chapter: IV. London, UK: Intech; 2012. pp. 85-106

[3] Loures L, Santos R, Panagopoulos T. Urban parks and Sustainable City planning-the case of Portimão, Portugal. WSEAS Transactions on Environment and Development. 2007;**3**(10):171-180

[4] Castanho RA. Sustainable urban planning in transboundary areas: Analysis of critical factors for territorial success [Doctoral Thesis]. Badajoz, Spain: University of Extremadura (UEx). Department of Vegetal Biology, Ecology and Earth Sciences; 2017

[5] Bilgili BC, Gökyer E. Urban green space system Plannings. In: Landscape Planning Chapter: IV. London, UK: Intech; 2012. pp. 107-122

[6] Jim CY, Lo AY, Byrne JA. Charting the green and climate-adaptive city. Landscape and Urban Planning. 2015;**138**:51-53

[7] Wolch JR, Byrne J, Newell JP. Urban green space, public health, and environmental justice: The challenge of making cities 'just green enough'. Landscape and Urban Planning. 2014;**125**(0):234-244

[8] Fadigas L. Urbanismo e Território—As Políticas Públicas. Lisboa, Portugal: Edições Sílabo; 2016

[9] Castanho RA, Cabezas J, Loures L, Fernández-Pozo L. Evolution of urban planning process in the Iberian Peninsula and his footprints in cityscape. New challenges. Revista Científica Monfragüe Resiliente. 2017;**VIII**(2):2340-5457

[10] Raposo M, Castanho RA, Machado M, Castro C, Santos P, Pinto-Gomes C. The relevance of vegetation series on the maintenance and sustainability of public spaces in the south-West Iberian Peninsula. In: Landscape Architecture: The Sense of Places, Models and Applications. London, UK: InTech; 2018. pp. 254-272

[11] Kaletová T, Loures L, Castanho RA, Aydin E, Gama J, Loures A, et al. Relevance of intermittent Rivers and streams in agricultural landscape and their impact on provided ecosystem services—A Mediterranean case study. International Journal of Environmental Research and Public Health. 2019, 2019;**16**:2693. DOI: 10.3390/ijerph16152693

[12] Spirn AW. The Eye Is a Door: Landscape, Photography, and the Art of Discovery. Cambridge, MA: Wolf Tree Press; 2014

[13] Delgado LE, Marín VH. Ecosystem services: Where on earth? Ecosystem Services. 2015;**14**:24-26

[14] Ibes DC. Integrating ecosystem services into urban park planning & design. Cities and the Environment. 2016;**9**(1):1. Available from: https://digitalcommons.lmu.edu/cate/vol9/iss1/1

[15] Glaves P, Egan D, Harrison K, Robinson R. Valuing ecosystem Services in the East of England. In: East of England Environment Forum. London, UK: East of England Regional Assembly and Government Office East England; 2009

[16] Nijnik M, Miller D. Valuation of ecosystem services: Paradox or

Pandora's box for decision-makers? One Ecosystem. 2017;**2**:e14808

[17] Breuste J, Qureshi S, Li J. Scaling down the ecosystem services at local level for urban parks of three megacities. Hercynian Forest. 2013;**46**:1-20

[18] Setälä H, Francini G, Allen JA, Jumpponen A, Hui N, Kotze DJ. Urban parks provide ecosystem services by retaining metals and nutrients in soils. Environmental Pollution. 2017;**231**:451-461

[19] Gratani L, Varone L, Bonito A. Carbon sequestration of four urban parks in Rome. Urban Forestry & Urban Greening. 2016;**19**:184-193

[20] Giedych R, Maksymiuk G. Specific features of parks and their impact on regulation and cultural ecosystem services provision in Warsaw, Poland. Sustainability. 2017;**9**:1-18. DOI: 10.3390/su9050792

[21] Sutton PC, Anderson SJ. Holistic valuation of urban ecosystem services in new York City's Central Park. Ecosystem Services. 2016;**19**:87-91

[22] EEA. EAGLE-RelatedProjects. 2018. Available from: https://land.copernicus.eu/eagle/files/eagle-related-projects/pt_clc-conversion-to-fao-lccs3_dec2010

[23] Global Cities Initiative. 2016. Available from: www.brookings.edu [Accessed on: 27 December 2019]

[24] Naranjo Gómez JM, Loures L, Castanho RA, Cabezas J, Fernández-Pozo L, Lousada S, et al. Assessing Land Use Changes in European Territories: A Retrospective Study from 1990 to 2012 [Working Title]. London, UK: IntechOpen; 2018. DOI: 10.5772/intechopen.80827

[25] Loures L, Panagopoulos T, Burley J. Assessing user preferences on post-industrial redevelopment. Environment and Planning. B, Planning & Design. 2016;**43**(5):871-892

[26] Loures L, Panagopoulos T. Reclamation of derelict industrial land in Portugal - greening is not enough. International Journal of Sustainable Development and Planning. 2010;**5**(4):343-350

[27] Loures L. Post-industrial landscapes as drivers for urban redevelopment: Public versus expert perspectives towards the benefits and barriers of the reuse of post-industrial sites in urban areas. Habitat International. 2015;**45**(2):72-81

[28] Loures L, Santos, R. e Panagopoulos T. Urban parks and sustainable city planning - the case of Portimão, Portugal. WSEAS Transactions on Environment and Development. 2007;**3**(10):171-180

[29] Loures, L. e Panagopoulos T. From derelict industrial areas towards multifunctional landscapes and urban renaissance. WSEAS Transactions on Environment and Development. 2007a;**3**(10):181-188

[30] Loures L. Industrial heritage: The past in the future of the city. WSEAS Transactions on Environment and Development. 2008a;**4**(9):784-793

[31] Loures L, Loures A, Nunes J, Panagopulos T. The green revolution – Converting post- industrial sites into urban parks – A case study analysis. International Journal of Energy and Environment. 2015;**9**:262-266

[32] Loures L, Panagopoulos T. Recovering derelict industrial landscapes in Portugal: Past interventions and future perspectives. In: Proceedings of the Int. Conf. On Energy, Environment, Ecosystems and Sustainable Development, Agios Nikolaos, Greece, July 24-26. 2007b, 2007

Chapter 10

Sustainable Utilization of the Lake Kinneret and Its Watershed Ecosystems: A Review

Moshe Gophen

Abstract

Lake Kinneret and its watershed have undergone significant structural modifications. Some of them are anthropogenic, and others are natural. Among natural modifications, the climate change is prominent. Among man-made changes, the drainage of the Hula Valley, construction of the National Water Carrier designating Lake Kinneret as a national resource of drinking water, construction of the Dam on the lake south outlet, agricultural developments in the Hula Valley, and others are included. Additional factors that recently have a significant impact on the management of the lake ecosystem are the development of ocean water desalinization capabilities. Nevertheless, these two ecosystems were carefully studied directed by limnological and wetland-agricultural scientific trait. One parameter was not yet accounted intensively, the assurance of long-term sustainability. This paper provides an overviewed insight into the conducted principles of management toward sustainable management after the establishment of the modifications as an attempt of predictive development outline.

Keywords: Kinneret, watershed, land use, climate change, sustainability

1. Introduction

The Lake Kinneret and its Drainage Basin had been widely investigated [1–6] aimed at enhancing basic scientific knowledge as well as ensuring submission of management recommendation to responsible national authorities. Vast majority of the studies were aimed at ad-hoc tracking ecological and anthropogenic changes. Insufficient attention was given to predictive design of systems sustainability. Anthropogenic management interventions are aimed at both, protection, and sustainable maintenance of the ecological structure and services. Nevertheless, ecosystem structures highly depend on the utilization trait. If a natural structure of an ecosystem is not threatened by anthropogenic intervention and/or by natural constrains, protection is smoothly implemented. Global climate change, increase of human population density, and consequently increasing demands for food production while tackling acute water scarcity became critical over larger global zones. Global promotion of desertification (decline of soil fertility) and afforestation amplified human intervention in natural ecological structure. Therefore, ecological management became a critical parameter which has a significant impact on human society. The ecosystems of Lake Kinneret and its watershed are subject for national and international concerns as sources for water supply, agricultural development, commercial fishery and aquaculture, territorial land for living (population dispersal), and more. During the last 80 years, natural and anthropogenic modification were accurately carried out over these ecosystems and consequent management operations were implemented aimed at efficient utilization of the natural resources of land and water while ecosystems sustainability will be preserved. A survey of the history and implementation of achievements which accompanied the ecological modifications in lake Kinneret and its watershed to ensure undamaged and sustainable ecosystem functionality is presented in this paper.

2. Background

2.1 The Lake Kinneret watershed

Lake Kinneret watershed is part of the Northern segment of the Syrian-African great rift Valley. The Lake Kinneret Watershed area (2730 km^2, of which 73% is an Israeli territory), from Kinarot Valley in the south to Upper Galilee (northeastern Israel) and southern Anti-Lebanon in Lebanon is 110 km long [1–4]. The Total area of the Kinneret drainage basin is 2730 km^2. It is divided into sub- units: (1) *Northern*: The River Jordan drainage; (2) *Eastern*: The southern part of the Golan Heights; (3) *Western*: The drainage area of Tzalmon and Amud rivers; (4) *South-Eastern and South-Western* minor sections. Versatile vegetations cover, soil and geological formations characterized the Kinneret watershed. Surface mean slope of the Kinneret Watershed is 2.8%. The following major events during the Anthropocene period were: The Lake Kinneret South Dam construction; The drainage of Old Lake Hula and swamps resulted a change of the regional Hydrological conditions; Regional population emigration and immigration; Governmental resolution of Lake Kinneret as major source for water supply and the construction of the National Water Carrier (NWC). These achievements established the long-term national policy of land use and water supply and geo-political boundary [3].

2.2 Regional hydrology

Three major headwater rivers (Hatzbani, Banyas and Dan) flow southerly downstream from the Hermon mountain region [2, 3, 5–11]. The Hula drainage changed the hydrological conditions: Jordan river crossing the Hula Valley splitting into two canals which joint at the south end of the Hula Valley flowing southerly downstream into Lake Kinneret maintains its Water Level (WL). Long-term (1926-present) record of WL daily monitor indicates maximal amplitude of 6.67 m (208.20–214.87 MBSL). The upper limited legislation of WL (208.8 mbvsl) was aimed at prevention of damage to previously constructed housing. The lower limit is flexibly affected by the location of the intake of the NWC (215 mbsl) and precaution of water quality impact. Since 1972 the hydrological management of the entire headwater resources was achieved by precipitation range, national water demands and NWC capacities and obviously by the south dam operation. [3]. Maximal lake water storage was achieved by close Dam limited by WL altitude. These were the management rules when 60% of national domestic water supply were originated from the Kinneret. Nevertheless, ecosystem sustainability aimed at water quality, mainly salinity, might be threatened if dam is closed and major withdraw is done through NWC [1–3, 5, 12]. Before Dam construction nutrient rich winter inflows crossed the lake through the upper water layers due to their higher temperature than that of the Epilimnion and naturally flew out through an open outlet. After Dam construction (1933) the outflow became human controlled aimed at water storing, and enhancement of salt and pollutants removal [1–3, 5, 12]. The final decision was a combination of actual conditions: precipitation-discharge range, desalinized water volume availabilities and lake water quality.

2.3 Land use

The territorial part of Israel within the Kinneret watershed is 2000 km^2 which is 73% of the total of 2730 km^2. The agricultural land use in the Kinneret watershed area is given in **Table 1** [3, 7, 13–15].

2.4 Water consumption

The total legislated water consumption for agriculture and domestic usage (source: National water authority) indicates the following: Until late 1990's it was ranged between 100 and 120 mcm (10^6 m^3) per year of which 99% for agricultural irrigation: 42% -grooves' 48% - field crops, 7% -fish ponds, and 1% - human domestic supply. Later on, restriction was instructed to 85 mcm/y and further to 68 mcm/y were implemented with additional supply from Lake Kinneret to the Golan Heights of 19 mcm/y [8, 9] (**Tables 2** and **3**).

How does agricultural management accept such constraints of natural drought and the followed legislation of water supply restriction? The answer was given in [15]: During 20 years (1990–2010) the efficiency of water utilization aimed at the beneficial revenue of agricultural production increased from 41,100 to 81,420 US$ per ha. It was the result of improvement of agricultural technology.

2.5 Land use land cover modifications in the Hula Valley

Before Hula Drainage the Valley was mostly (6500 ha) covered by natural wetlands (Peat soil) and old Lake Hula (1300 ha), Hula drainage, converted natural wetland into agricultural land [7, 15, 17, 19]. It was an infrastructure development for an agricultural income source for the local immigrated residents. Between 1960 and 1990 the Peat-Land area cultivation has yielded economically sufficient products. Nevertheless, contributed nutrient to Lake Kinneret threatened its water quality. It was resulted by inappropriate irrigation methods. The outcome was peat soil destruction and subsidence, dust storms which blocked the drainage canals, underground fires and rodent population outbreaks. Agricultural crops were damaged and Kineret water quality became threatened. A reclamation

Type of land cover	Area (km$^{2)}$
Field Crops	180
Orchards	197
Fishponds, reservoirs, Agmon, Lake Kinneret	171
Natural Forest and Grove	266
Not Cultivated land	1067
Other	111
Total	1992

Table 1.
Israeli agricultural land use (km^2) in the Kinneret watershed as documented in 2004.

Geographical region	Surface area (km^2)	Annual rainfall (mm/y)	Annual rain volume (mcm/y)
Eastern-Northern Galilee	542	800	434
Jordan-Hermon	788	900	709
Hula Valley	200	450	90
Golan Height	580	900	522
Western Basin	450	450	202
Small Southern Basins	170	450	77
Total	2730	(Mean: 658)	2034

Table 2.
Precipitation Regime: Geographical sub-units of the Lake Kinneret drainage basin, their surface area (Km^2), annual precipitation gauge (mm/y) and calculated total rainfall volume (mcm/y) are given [8, 9, 13, 14, 16, 17].

Used-cover type	1949	1958	1976	1986	2010
Water	24%	0	0	2%	2%
Swamps	54%	7%	7%	3%	7%
Flooded	22%	0	0	0	0
Field Crops	0	59%	79%	58%	68%
Uncultivated	0	17%	—	14%	5%
Other	0	8.5%	3%	10%	7%
Orchards	0	0	3%	8%	9%
Fish ponds	0	8.5%	8%	5%	2%

Table 3.
The history of land use/land cover included in the peat soil convention (PSC) in the Hula Valley (59 km^2; 5900 ha). Numbers are % of the total area. Historical events: 1952–1957 drainage and conversion to agricultural management; 1989–1995—Hula reclamation project (HRP) implementation [18, 19].

project (Hula Reclamation Project, HRP) was consequently discussed and implemented. A shift of 500 ha Peat-Land from agriculture to eco-tourism usage was achieved. The HRP concept was aimed at ecosystem sustainability and therefore based on anthropogenic intervention combined with the introduction of natural plants. Reconstruction of the hydrological drainage system of the entire valley was renovated. The critical need for soil structure protection by maintenance of its moisture was achieved by implementation of irrigation method of moveable sprinkle line [14, 17].

2.6 The peat soil convention: Sustainable achievement

The major significant variable of regional water balance is obviously rainfall contribution. Although a major part of the regional rainy waters input is transformed into runoff, flowing downstream into Lake Kinneret, significant volume of is migrated into unknown underground spaces in the Hula Valley. When climate is changed, and therefore, water consumption and possibly land use policy reduces, it will have an impact on lake water level. The second level of water consumption is due to Evapo-transpiration (ET). This variable of the regional water balance is strongly affected by climate conditions, land plant cover, water availability and soil properties. Dryness conditions enhance soil moisture reduction, which is affected by land use policy of slow down crops cultivation (plant coverage restriction) and, therefore, reduce regional evaporation capacity. The management of the Kinneret watershed is a good example of protection of sustainability of an ecological ecosystem where natural and anthropogenic interests are together combating dryness in the hula Valley (**Figure 1**) [7, 13–15, 17].

As a result of enhanced dryness and water supply limitations, the national policy of water pricing was reordered. Consequently, cost account of water consumption in the Hula Valley became more expensive. Nevertheless, as part of the National Water Authority recognition of ecosystem sustainable management, a formal confirmation was carried out of the special status of the Hula valley. Followed by legislated water price reduction accompanied by stakeholders' commitment to irrigate fields despite being bare in summer. Nonprofitable expenses were compensated by the lowering of water pricing. The difference between the National and the reduced tariff was dedicated to a stakeholder's managerial foundation to cover those irrigation expenses.

Results in **Table 4** indicate a reduction of Water-Swampy-Flooded area from 100% to less than 5% surface cover. As a result of enhanced dryness (water scarcity)

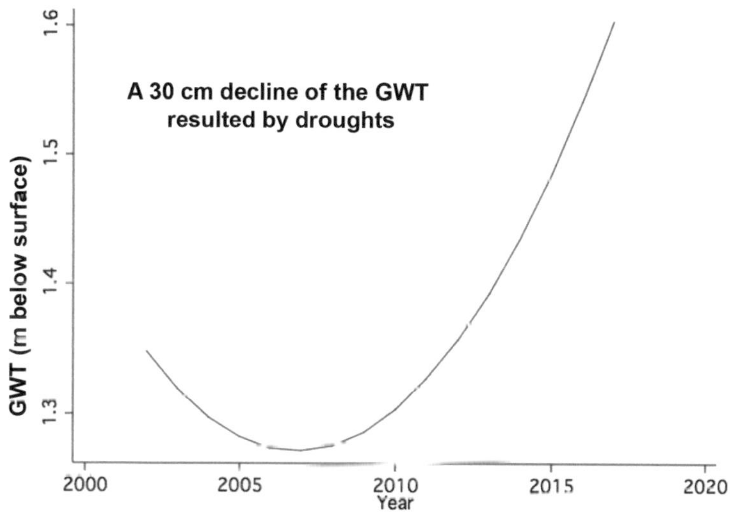

Figure 1.
Annual Total Hula Valley region average of ground water table (GWT) (m below surface).

WL range (mbsl)	Number of months (%)
Below 214	32 (6)
214–213	65 (11)
213–212	89 (15)
212–211	104 (18)
211–210	144 (24)
210–209	125 (21)
Above 209	30 (5)

Table 4.
Number of months with monthly means of WL with respect to 1 m WL interval in Lake Kinneret (1970–2018).

driven by climate change during the recent 15 years field crops area in the watershed was restricted by 35% and Fishponds by 43%. Although agricultural land-use in the Watershed was reduced as well as water availability (from 110 mcm/y to 68 mcm/y) crops and revenue per areal unit were improved simultaneously. This was resulted by technological improvements and land beneficial significance. In other words, natural constrains of water scarcity were achieved by water and land utilization efficiency aimed at sustainability maintenance.

3. Climate change

3.1 Precipitation and discharges

Since mid-1980's precipitation decline in the Kinneret Drainage Basin was documented (**Figures 2–6**). During 2013/14 and 2015/16 seasons rainfall was 47% and 68% respectively below the multiannual mean. Major contributors to the Jordan discharge are Dan and Banias rivers. The discharge of Dan and Banias during 2014 (2.67 and 0.16 m^3/s respectively) were the lowest since recent 22 years in comparison with the maximum discharges of 12.8 and 7.4 m^3/s respectively [8, 9]. The annual discharges of those rivers declined by 63 and 14 mcm/y respectively. As a result, annual availability of lake water (inflow minus evaporation) during 1985–2016 indicates a decline from 470 to 225 mcm/y. As the result of promoted trend of dryness, the hydrological dynamics of the Lake Kinneret ecosystem was modified. The Input reduction accompanied by water level decline and elimination of pumping together with close Dam policy eliminated exchange level and prolongation of RT from 5 to 7 to a range of 15–>20 years. Evaluation of SPI (Standard Precipitation Index) values from 87 years precipitation record has indicated 11 and 17 negative indexes (aridity level) during1927–1970 and 1970–2014 respectively, which is an indication of climate change toward dryness. River discharge reduction initiated also changes of the phytoplankton community structure in Lake Kinneret. The Nitrogen supply was diminished resulted Peridinium decline which was replaced by Cyanobacteria dominance.

3.2 Rain and headwater discharges

Decline of rainfall and Jordan River discharges during the last 40 years and historical deficiency of aquifers storage in the Israeli Northern Basin for 100 years were documented by the Israel Hydrological Service [1, 5, 8, 9]. During 2013/14 and 2015/16 rainfall was 47 and 68% respectively below the multiannual average. The Rivers Dan and Banias discharge during 2014 (2.67 and 0.16 m^3/s respectively) were the lowest since recent 22 years. Major contributors to the Jordan flow are Dan and Banias rivers. The annual discharges in those rivers declined by 63 and 14 mcm/y, respectively. A decline from 470 to 225 mcm/y of availability of Kinneret waters (mcm/y) was indicated during 1985–2016.

3.3 Air temperature

The records of daily Maximum and Minimum of air temperatures measured at the Meteorological Station Dafna (northern part of the Hula Valley) indicates an increase since mid-1980s. The air temperature record indicates [13, 14, 17, 19] an annual maximum and minimum elevation by 2.7 and 1.5°C, respectively. Studies on regional water balances confirmed enhancement of water loss not only as

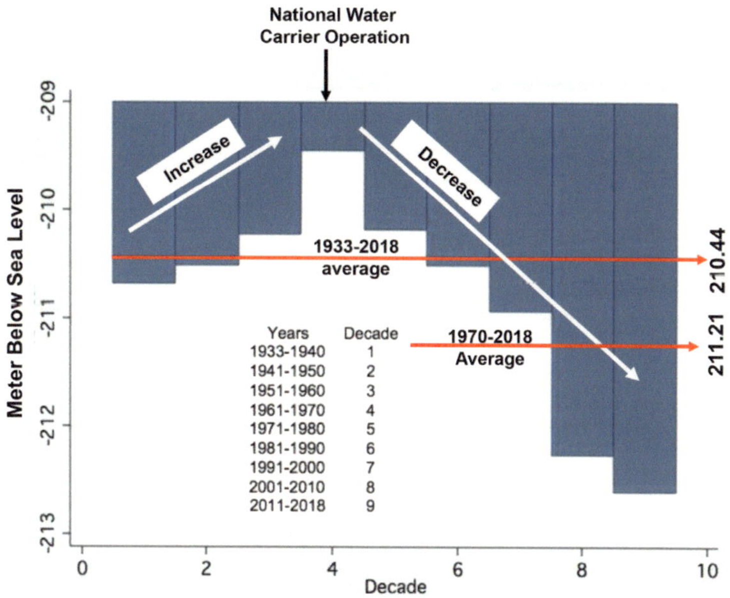

Figure 2.
Groups of decade (10 years) averages of monthly means of water level in Lake Kinneret. Trend of changes, periodical means, and anthropogenic events are indicated.

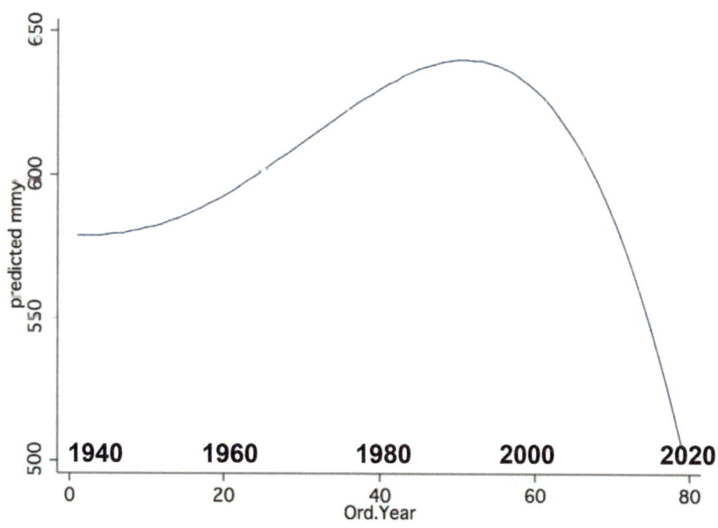

Figure 3.
Fractional polynomial regression between annual (1940–2018) precipitation and years.

precipitation and runoffs but also in the underground preferential cavities in the peat soil. Dryness processes of the Hula Valley soil confirm the potential loss of water during dryness periods. Therefore, it was recommended to prevent decline level of soil moisture by suitable irrigation method. Recommended Optimal management is, therefore, moisture enhancement especially during summer time. Climate change and consequent dryness constrains initiated a special legislation, the Peat Soil Convention, which ensured summer water supply for irrigation.

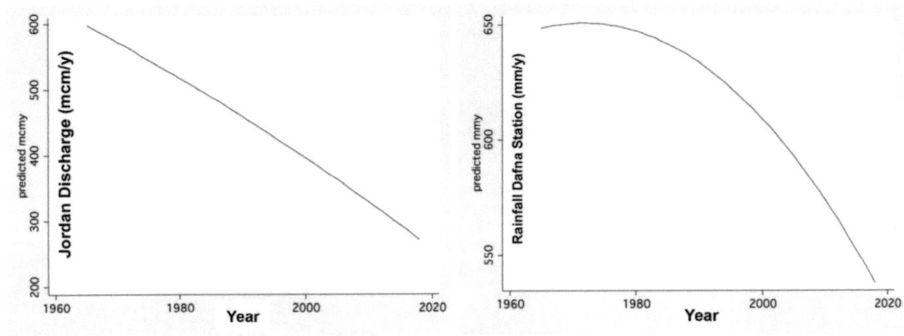

Figure 4.
Fractional polynomial regression between annual Jordan discharge (mcm/y) (left panel) and rainfall (mm/y) (Dafna Station, right panel) during 1969–2018.

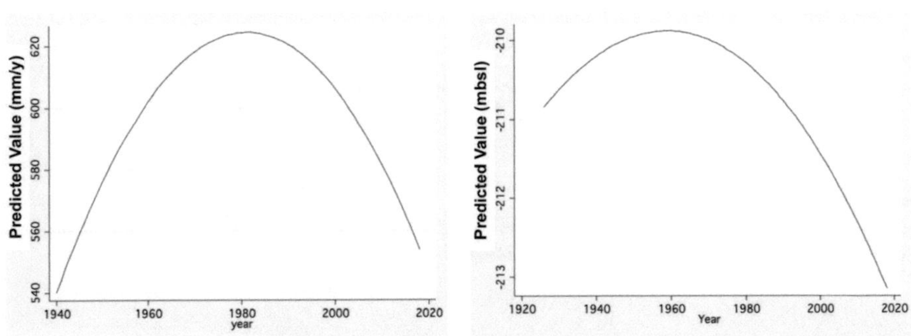

Figure 5.
Fractional polynomial regressions between rainfall (left panel, Dafna Station), and annual means of WL (mbsl) (right panel), and years.

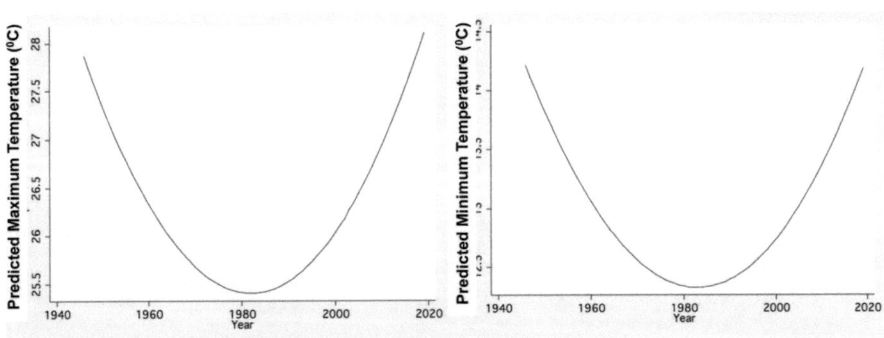

Figure 6.
Fractional polynomial regressions between annual means of daily maximum (left panel) and minimum (right panel) air temperature measured in Dafna Station and years during 1940-2020.

4. Lake Kinneret ecosystem

4.1 Water level fluctuations in Lake Kinneret

The obvious direct relation between Kinneret WL and precipitation regime in the watershed was documented widely in previous studies. Historical (9000 years before present) data of the Kinneret WL was investigated by two different methods [20, 21] and indicated an amplitude of 20 m (197–217 mbsl) WL fluctuations.

Monthly means of daily WL measurements has indicated that during 48 years only 97 months (17%) WL was lower than the legislated bottom line of 213 mbsl occurred mostly during recent 18 years (**Table 4**) [1, 3, 10, 16, 18–23] .

Results in **Table 4** and **Figure 2** indicate that during most of the time (83%) the Kinneret WL was not lower than 213 mbsl and, before 2000th, WL was higher than the minimal legislated altitude. The decline of WL below the instructed WL bottom- line (213 mbsl) was recorded during years of exceptional decline of rainfall: 2000–2002, 2008–2011, and 2016–2018, which consequently resulted in significant restriction of agricultural water allocation by the National Water Authority.

4.2 Nutrient dynamics

The discussion about dependence relations between phytoplankton and nutrients presented here emphasize the paradigm of an everlasting dilemma: Between phytoplankton composition and nutrient concentrations, who is the boss? (**Figures** 7 and **8**) Algal community structure responds to the concentration of the nutrients or the contrary: does the nutrient concentrations is primary or secondary result of the algal density [6, 11, 23–33]? Nitrogen input is defined as predictor of algal domination in Lake Kinneret: Peridinium or Cyanobacteria. A decline of epilimnetic TN standing stock was documented during 1969 to 2001 accompanied by decline of Peridinium biomass while the biomass of Cyanobacteria increased. TN decline initiated Nitrogen deficiency, which is favored by Cyanobacteria [32] due to their ability of maintain the fixation of atmospheric Nitrogen by the enzyme of Nitrogenase. Earlier studies suggested two elements as key factors for the Peridinium bloom formation: Copper (Cu) and Selenium (Se) [6, 23, 29–33]. The study of the Cu impact was not thoroughly developed but that of Se did it thoroughly. It was confirmed that Se is a limiting factor of Peridinium growth. Before Hula drainage, the chemical conditions of the peat soil were mostly reductive but presently more oxidative and, therefore, limitation of Se is not impossible. Earlier Studies [29, 30], suggested that precipitation and runoff discharges are an important source of bioavailable Se (Selenites and Organic Se) and high availability of Se in surface waters of Kinneret watershed might be a significant supporter of the Peridinium heavy blooms. Therefore, it is suggested that in addition to Nitrogen deficiency, Se input decline affected the decline of Peridinium biomass. Conclusively, the replacement of Peridinium by Cyanobacteria is mostly due to change of nutrients dynamic resulted by climate change. The depletion of Nitrogen supply is based on a long-term record and the recorded data about Se dynamics is partial.

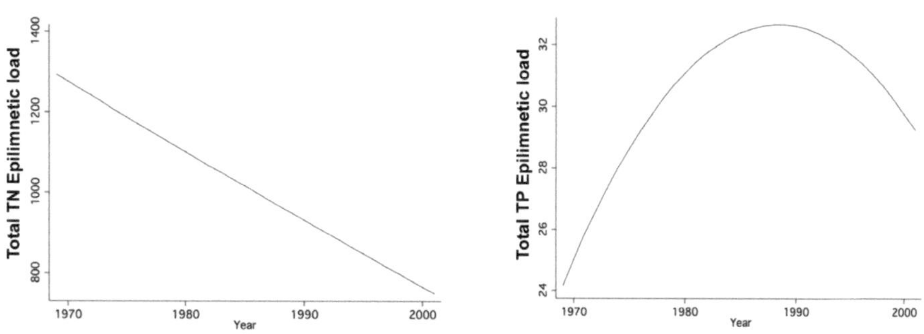

Figure 7.
Fractional polynomial regressions between annual averages of Epilimnetic loads (ton) of Total nitrogen (TN) (left), Total phosphorus (TP) (right) inputs through Jordan inflow and years.

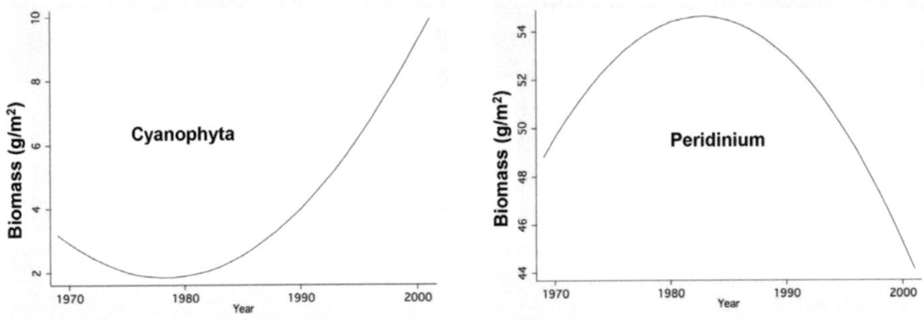

Figure 8.
Fractional polynomial regressions between annual averages of phytoplankton biomass (g/m²) (Peridinium, Cyanophyta) and years.

4.3 Dam management: to open or not to open?

Open or Not to Open (ONO) the South Dam when WL is high? That is the question for Sustainability by hydrological managers [23, 34, 35]. Regional trends of climate change and dryness process were recorded: Standard Precipitation Index (SPI) enhancement precipitation decline, air and lake water temperature increase, river discharges and restriction of lake input volumes and consequent decline of WL, elongation of RT duration. The decline of water availability for domestic and agricultural supply created a national concern accompanied by increase of Lake water salinity, epilimnetic Nitrogen deficiency and Phosphorus sufficiency which enhanced biomass replacement of Peridinium by Cyanobacteria [28]. These natural ecological modifications were accompanied since 2010 by replacement of the lake as principle supplier of drinking water by desalinization of mediterranean waters. The following additional parameters made the ONO dilemma more significant. Multi-annual (1933–2020) daily record of WL indicates an average annual increase of 1.6 m. Nevertheless several annual exceptions of higher and/or lower of it are common. These exceptions are critical for decision makers with regard to the dynamics and management policy of water supply which was dependent of pumping rate and Dam management: high WL indicate pumping potential enhancement and low WL dictate withdraw restrictions. Several cases which represent not common conditions are: During fall 2001 WL was lowered to the lowest altitude ever recorded since 1933–214.87 mbsl and pumping was exceptionally restricted; during winter 1969 WL increased up to 208.2 mbsl and the dam was maximally opened; Five hydrological years (October–September next year) 2013/2014–2017/2018 were a drought sequence in a row when the annual increase of the WL varied between 0.35 and 1.58 m. At the end of this drought period the epilimnetic salinity was 325 ppm Chloride which was even predicted to increase higher if dryness trend would be continued. Three years earlier (2011–2013) the WL annual elevation varied between 1.75 and 2.58 m. After five drought seasons (2014–2018) heavy rain winters came and WL elevation was 3.41 in 2019 and 3.0 m in 2020. In December 2019 when WL was 211.89 mbsl, salinity was measured as 325 ppm. chloride. Later on in winter 2019 the heavy input discharges during January – mid-March when dam was closed dilution effect resulted salinity decline to 273 ppm chloride, (52 ppm decline). It is likely that enhanced water exchange (RT shortening) by open dam might cause higher decline of salt concentration. Moreover, it is also predicted to enhance nutrients and Microcystis biomass removal which enhance improvement of water quality. Since late 1990's the phytoplankton assemblages are dominated by Cyanobacteria, mostly due to the toxic Microcystis spp. The recent lake situation is therefore creating a dilemma for future management of sustainability: Water supply is done by desalinization, while salinity and Microcystis are enhanced supported by close dam and RT elongation water quality is therefore deteriorated. It is likely that, within future design for sustainability other than hydrological factors must be included. For example, salinity, nutrients and toxic Cyanobacteria biomass Consequently, during rainy winter a partial open of the dam is recommended aimed at quality improvement.

4.4 Salinity

The salinity of Lake Kinneret water was a critical parameter when supply for agricultural utilization was actually required. The major supply of salt to the lake are fluxed through the lake bottom through two major process: surface infiltration (superficial) and welling up. Salts' contribution through rivers and tributary inflows are much lower in comparison the sub-lacustrine sources. The salinity of River Jordan (65% of total inflows) is more than 10 times lower than that of the lake. Nevertheless, until late 1950's about 25% (total about 160000 tons annually) of salt input came through the runoff of two hot-salty springs located close to the north-western lake shoreline. Those two springs were diverted (1967) and about 40,000 tons of salt were eliminated from the lake budget. As a result of this anthropogenic implementation accompanied by the heavy floods during the winter of 1968–1969 (25% of lake water were exchanged) lake water salinity declined from 400 to 210 ppm Chloride. Historical information indicates Chloride concentration range before the 1950's between 290 and 325 ppm. A critical question is therefore arise: why salinity was increased during 1948–1968 from 280 to 400 ppm Chloride when negligible consumption of lake water was supplied for domestic and agricultural usage and the only one management tool, Dam operation (NWC was not in use yet) was available? The WL record indicates an increase of more than 2 m during 20 years (1948–1968). It is therefore suggested that Dam operation policy was aimed at long-term water accumulation causing WL elevation accompanied by Salt accumulation which resulted significant concentration increase of Chloride by more than 100 ppm. It has to be noted that during 1948–1968 there was also a sequence of several drought seasons in a row. Conclusively, for 20 years, the water exchange was low resulting elongation of Residence Time duration. The 1948–1968 event is a case study example for future consideration of sustainability design. The case study of 1948–1968 was not the only one for future consideration. Two other closely related cases which continued only one winter each are relevant: During two winters with heavy rain, in 1968/69 and 1991/92 similar inputs of 1×10^9 m^3 were fluxed into the lake during 2 months. The difference between the two winters was the Dam operation [23, 34]: During the first winter the Dam was maximal open and during the second winter completely closed. It has to be considered that the input of low salinity river waters into the much higher salt concentration of lake water create a dilution effect in winter and salt concentration in the lake decline but the level of the decline is dependent on two parameters: input volume and water replacement dynamic and the level of replacement is the dependent of open Dam policy. Results indicated that open Dam operation enhanced water replacement (exchange) which is in fact Residence Time shortening. Therefore during the winter of 1968/69 the Chloride concentration declined by 64 ppm while in the winter of 1991/92 the decline was smaller – 39 ppm. It is therefore recommended to enhance water exchange (shorter Residence Time) through open Dam or pumping regime to remove salt and other pollutants (including biomass of Cyanobacteria) if water storage for supply is not critical resulted by desalinization supply.

4.5 Residence time (RT) prolongation

Residence Time prolongation (Water exchange reduction) affected also Nitrogen and Phosphorus dynamics (**Figures 9–11**). The process of Peridinium decline and Cyanobacteria enhancement was also supported by RT prolongation. The climate change initiated a linkage of chain events. Discharge decline and water scarcity (dryness) resulted WL decrease, RT prolongation and nutrient supply reduction accompanied by the modification of algal community structure. Normally, the higher the discharge is the faster is the increase of the WL and the shorter is the RT. and vice versa. The decline of discharge and Nitrogen input was accompanied by decline of epilimnetic TN stock and decrease of TN/TP mass ratio. Optimal Ecosystem management is aimed at protecting sustainability and the operation tool is through hydrological control: desirable ranges of pumping, WL fluctuations, nutrient dynamics preventing water quality deterioration resulting adequate water quality. Before 2010 the majority of domestic water supply originated from Lake Kinneret but essential climate change condition constrains created the need for the construction of alternative water source - Desalinization. The decline of discharge and insufficient Nitrogen input caused the phytoplankton community change. The newly created ecosystem structure enforced management adaptation for sustainability protection. When water budget is positive accompanied by appropriate withdraw (pumping and/or open Dam options) RT become shorter and water exchange is high. Decline of Nitrogen availability accompanied by Phosphorus enhancement caused the decline of TN/TP mass ratio [23, 35]. Hydrological management of Lake Kinneret is creating a dilemma for future implementation: Water supply is done by desalinization, salinity and Microcystis are enhanced as supported by close dams that enhance RT prolongation and water quality deterioration. It is therefore likely that recently, WL regime is not the management key factors and other parameters should step forward on the scale of priorities such as: salinity, nutrients and toxic Cyanobacteria biomass. For example, during heavy rain a partial open of the Dam is recommended to remove salt, phosphorus and Cyanobacteria biomass while water supply is not critical.

4.6 The impact of residence time on nutrients and phytoplankton dynamics

Monthly changes of epilimnetic TN stock were found to be related to the length of Residence Time (RT; the ratio between inflows rate and Lake Volume): higher TN stock accompanied longer RT [23, 27, 28]. Lake Volume increase and shorter RT are correlated with the epilimnetic load decline of TN an increase of epilimnetic TP loads during January–April and gradual decline later on until December was correlated with the Hydrological parameters: RT elongation during January–September, became shorter later; The decline of TN/TP Mass ratio is respective to RT prolongation: The higher the RT value is, the lower is the Epilimnetic TN/TP mass ratio [24, 26, 27]. The biomass of Peridinium contributes Phosphorus and the headwater input are carriers of Nitrogen. The shortest RTs were recorded during the Peridinium bloom onset and later when RT length declines, P-mediated Peridinium dissipates, and Epilimnetic stock diminishes. Shortest RT's were recorded during winter, and later in the year RT becomes longer. Conclusively, external hydrology (water discharge) contribute Nitrogen and Peridinium bloom in addition to dust deposition, external inputs, and bottom sediments by microbial activity contribute Phosphorus. When Nitrogen supply was declined Peridinium bloom was deleted and Phosphorus fluxes were shortened and TN/TP mass ratio was lowered. The final result was enhancement of Peridinium replacement by Cyanobacteria [28].

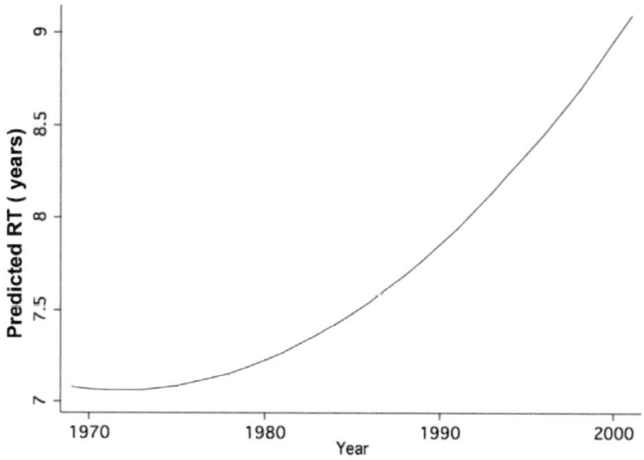

Figure 9.
Fractional polynomial regression between annual means of monthly residence time (RT in years) and years.

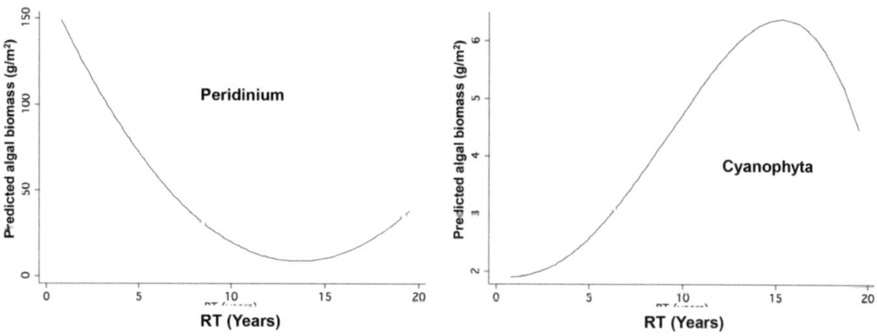

Figure 10.
Fractional polynomial regression between RT length (years) and algal (Peridinium, Cyanophyta) biomass (g/m^2).

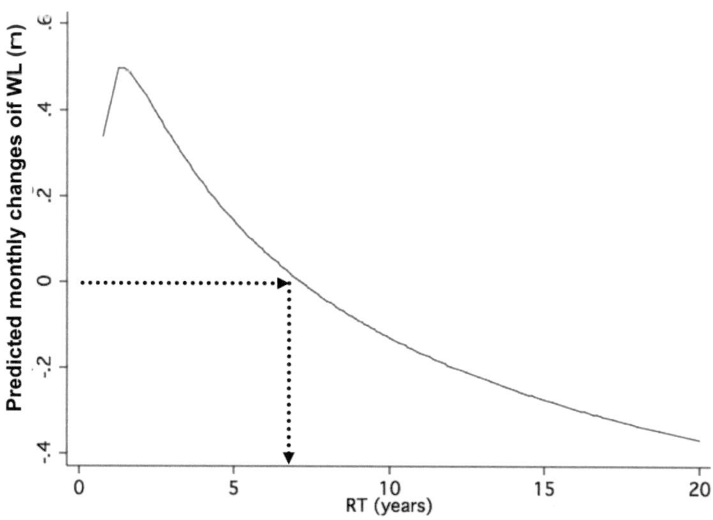

Figure 11.
Fractional polynomial regression between RT length (years) and monthly changes of water level (m).

4.6.1 Peridinium

Elongation of RT corresponds to the reduction of the Peridinium biomass. The RT elongation is a signal of Nitrogen availability deficiency. A slight increase of Peridinium biomass documented during the longest RT, is probably attributed to Nitrogen input by fixation carried out by Cyanobacteria.

4.6.2 Cyanobacteria

A prominent increase of the Cyanophyta Biomass (from 1.9 to 6.3 g/m^2) was documented in response to RT elongation of 1 to 15 years accompanied by decline of Nitrogen availability. It is likely that the Nitrogen deficiency in lake water was compensated by Nitrogen fixation maintained by Cyanobacteria. It is assumed that the minor decline of Cyanobacteria biomass observed during the longest RT is due to the lack of Phosphorus when Peridinium is absent.

5. Fishery

The Fishery management in Lake Kinneret is aimed at both, commercial income and water quality protection and ecosystem sustainability [34]. As a result, stocking of exotic fish species was confirmed just of those which cannot reproduce in the lake, their feeding habit improve water quality and their contribution to commercial fishery is essential. Final confirmation was given after a thorough investigation which confirm the implementation of those three objectives. The Tilapia *Sarotherodon galilaeus* was indicated as an optimal species target: the species is native, feed intensively on the bloom forming Peridinium and have a high commercial value. Therefore, fishing efforts are mostly aimed at this fish and the lake population is enhanced by commercial fingerlings production. Results in **Table 5** summarized annual landings of *S. galilaeus* [36–40].

Respective data of other stocked species indicates the followings: The stocking of *Oreochromis aureus* which is not pure native species in the lake was eliminated due to food competition with preferred *S. galilaeus*; Until the mid-1990s, stocking of Silver Carp (*Hypophthalmichthys molitrix*) was not recommended aimed at enhancement of zooplanktonic algal grazers whereas later on when Microcystis replaced Peridinium its stocking was recommended due to its efficient consumption of this algae. Three Gray mullet species (Marine origin) are successfully stocked because of ecological adaptation to improve water quality, not able to reproduce in Lake Kinneret and has high commercial value. Another 7 other species of exotic species were totally deleted from stocking program. Conclusively, stocking resources are invested toward fish species that has positive impact on water quality, fishermen income, and the exotics are unable to reproduce in the lake. The fishery (landing and stocking) management policy contribute strengthening of

Period	Trend of change	Periodical averaged landing (t/year) (SD)
1959–1970	Stable	175 (28)
1970–1990	Increase	248 (112)
1990–2010	Decline	231 (154)
2011–2016	Increase	184 (99)

Table 5.
Periodical means (SD) of Sarotherodon galilaeus landings (t/year) and indication of trend of changes [36, 37].

ecosystem sustainability. Peridinium was the major food source for *Sarotherodon galilaeus*. Several other constraints created additional pressures on the fish population: Increased population of the migratory fish predator, Great Cormorant (*Phalacrocorax carbo*), reduction of stocked *S. galilaeus* fingerlings, usage of illegal fishing gill-nets mesh size, the elimination of Bleaks (Sardine: *Mirograx terraesanctae terraesanctae, Acanthobrama lissneri*) fishing, enhanced piscivory of *S. galilaeus* by *Clarias gariepinus* and outburst of Viral diseases, which infected mostly Tilapias. Ecological structure with complicated interactions require informative record long enough to ensure appropriate management decision in response to actual and unusual developed changes. Inappropriate alerted conclusions were followed a fishery crisis in Lake Kinneret when annual landings of *S. galilaeus* in 2007–2008 were less than 10 tons while normally its varied between 100 and 300. Simultaneously, documentation of the total number of fish (>90% Bleaks) was gradually increasing between 1987 and 2005. A recommendation of a three-year total fishing ban in Lake Kinneret was concluded. This decision was alternatively replaced by a recommendation of normal continuation of fishing. The fishing ban decision was canceled, and fishing continuation was confirmed formally. During 2010–2016, the population of *S. galilaeus* and consequently their landings were recovered and came to its normal level. During 2007–2008, Tilapia fishery in Lake Kinneret collapsed [18, 19]. A governmental decision of 3 years total commercial fishing ban was undertaken. Nevertheless, as part of ecological sustainability clarification of potential reasons the resolution was canceled and within 3 years the *T. galilaeus* population recovered. [36, 39]. The changes of the Phytoplankton composition were also accompanied by a modification of the fish feeding habits. During its dominance, *Peridinium spp.* was the major food component of the most valued native fish (*Sarotherodon galilaeus*) in the lake. Zooplankton was the major food constituent of the endemic Bleak cyprinids (*Acanthobrama terraesanctae terraesanctae*, *Acanthobrama lissneri*). To ensure water quality, it is important to maintain high grazing pressure of zooplankton on nano-phytoplankton. Removal of the unwanted Bleaks by intensified fishery management and the introduction of the exotic Silver Carp *(Hyphophthalmichthis molitrix),* an efficient consumer of *Microcystis,* is therefore beneficial. Zooplankton biomass in Lake Kinneret declined from 1970 to the early 1990s but increased thereafter. Both, the biomass and size frequency of cladocerans were affected by fish predation. Under the modified food web structure, Tilapia became a competitor with Bleaks on Zooplankton consumption. Information given in previous studies including the long-term record of the Kinneret zooplankton [1–3, 6] distribution, population dynamics and physiological trait was re-evaluated in the present paper.

5.1 Zooplankton dynamics

The Zooplankton compartment within the Kinneret ecosystem exemplify the necessity for multi targeted maintenance evaluation [1, 2, 5, 6]. The complex interaction relationships require a comprehensive implementation. Long term (1969–2001) averages of zooplankton biomass (WW) density in Lake Kinneret is given in **Table 6** as averages and ranges (Max-Min) of annual means.

A deeper insight into the Zooplankton temporal distribution indicates long term decline since mid-1980s accompanied enhancement of Bleak populations. The Bleaks population increase was resulted by decline of fishing pressure. Therefore, a recommendation was submitted and accepted to subsidize Bleak fishing. The concept of sustainability included reduction of cascaded top-down pressure on algal grazers to improve water quality. Nevertheless, ecosystem sustainability protection requires a comprehensive approach of which only fishery was accounted. To achieve water quality improvement by algal biomass reduction in oligotrophic deep

Group	Average (g(ww)/m^2) (%)	Max-min range (g(ww)/m^2)
Copepoda	9.0 (33)	2.3–17.7
Cladoceraa	15.9 (59)	8.8–25.1
Rotifera	2.1 (8)	0.9–5.2
Total	27.0	12–48

Table 6.
Averages of annual (1969–2001) means and max-min ranges of zooplankton groups (Copepoda, Cladocera, Rotifera, Total) WW-biomass (g(ww)/m^2).

lakes Phosphorus removal is ultimately required. Because Phosphorus removal was excluded Sustainability protection was only a partial success: zooplankton biomass was recovered but algal biomass was not reduced [36–39]. The suppression of the enhanced population winter migratory fish consumer Cormorants in Lake Kinneret became essential as a protector of ecosystem sustainability [36–39]. The deportation of Cormorant from Lake Kinneret is a useful implementation of water quality protection. The number of Great Cormorant *(Phalcocorax carbo)* wintering (from the end of October through March) in the Lake Kinneret Region is approximated as 6000 (5000–7000). The predation rate of the Cormorants indicates a daily ration varying between 300 and 1000 grams per bird with the more common value of 700 grams per bird [37, 38]. Six thousand Great Cormorants preying daily at 500 g fish per bird during 100 days removed 300 tons of sub-commercial-sized Tilapia (Mostly *S. galilaeus*) from the lake. However, we have to take into account that the fishes preyed on are below the commercial size of 100 g per fish, that is to say that the potential damage is bigger (legal size >200 g/fish). Individual Tilapia preyed on weighted 50-70 g; if not preyed on they might grow up to commercial size within 5–6 months to be marketed. Consequently, the commercial value of such losses is between 1.5 and 3.0 million US$. Such a damage to fishermen's income and ecologically to the system can be reduced by aggressive deportation of the Cormorants from Lake Kinneret and simultaneously from their night station site. The ecological contribution of Tilapia to the ecosystem aimed at water quality protection is done through the consumption of *Peridinium* biomass gradually reappeared recently. The recommended accompanied operation is Bleaks removal aimed at releasing zooplankton food biomass to *S. galilaeus*. Predictive recommendations include, among others, is a practical design which is presently under consideration aimed at achieving reduction of fish predation by Cormorant without violating accepted legislations. In other words to protect nature items together with improvement of fishery and water quality in Lake Kinneret.

6. Shallows: beach vegetation interface

The lake shallows/beach interface is a contradiction between public and eco-limnological services. The surface area of the inundation zone is about 11 km^2 according to: Annual WL fluctuates between 209 with lake bottom area is 168.9 km^2, and 213 mbsl with lake bottom area of 161.4 km^2, lake shoreline length is 55 km and adjacent beach belt width is 50 m. This nearby water beach area is potentially open for recreation service entitled "Aquatic Recreation Belt" (ARB) [41]. Nevertheless, under temporal long-term inundation regime the ARB allocation is not precisely predictive. During heavy precipitation season WL is high and major part of the ABR area is shrunk while after low rainfall season ABR area is wider and immediately covered by beach aquatic vegetation. The fast grower aquatic plants create a nuisance for aquatic recreational activities such as water access and favored environmental conditions for unwanted animals like Venomous Snakes. Fox, Mongoose, Jackal, etc. Moreover, next year the aquatic plants would be flooded and decomposed forming optimal conditions for Mosquitoes reproduction accompanied by accumulation of rotten bad smell organic matters. Reasonable solution might be mowing of those plants which on the other hand probably create shortage of spawning ground for *S. galilaeus* [10]. The Kinneret shoreline length is 55 km of which only 12.7 km (23%) are legal open public beaches. So far, prognosis of damage is practically negligible while enhancing *S. galilaeus* population biomass is possible by commercial production of fingerlings. Conclusively, partial mowing of beach vegetation and *S. galilaeus* reproduction would not be interfered. These objectives are due to the high (212–213 mbsl) WL regime. A recent computation of lake water surface area in respect to WL obviously indicates close positive significant linear regression when WL was below 210 mbsl. Under higher WL the relation was insignificant. It is because WL came the Bethsaida lagoons altitude. Resulting lower elevation of WL with respect to wide flooding area. The Beteicha lagoons densely covered by aquatic plants (*Tamarix spp.*, *Typha spp.*, and *Phragmites spp.*) are known as an optimal spawning ground, YOY care treatment for *S. galilaeus*. Conclusively [10], beach vegetation mowing as a compromise between fish reproduction interference and human recreation is relevant when WL altitude is lower than 212 mbsl.

7. Hula valley farmers and Kinneret limnologists should be friends

Since 1993 flocks of migratory Cranes (*Grus grus*) stay during 4 winter months in the Hula Valley. The Crane wintering provided the most attractive target for Eco-tourism [42]. The winter migrating of app. 50,000 Cranes in the Hula Valley during 4 months are very attractive, and the touristic visits were enhanced significantly from about 50,000 during the early 1990's to almost half a million presently. The Crane wintering flocks created severe difficulties, including damage of agricultural crop and nutrient (excretions) sources in Lake Agmon-Hula and further downstream into Lake Kinneret. It might be risky for the stability of the Kinneret Sustainable trait: 50 × 10^3 Cranes excrete 5.24 gP/Ind./day during 170 days produce approximately 44.5 tons of TP [42] beside other TP sources in the Hula Peat soil, agricultural fertilization and ecological processes in Lake Agmon.

Protection of aquatic Ecosystem sustainability require anthropogenic control throughout the entire watershed. The social, agricultural, hydrological and ecological activities of development in the Hula Valley justify a careful approach., The Crane case, among others, require a significant consideration. The Hula Valley contribute above 50% of the external nutrient inputs into Lake Kinneret and the agricultural management has an impact on nutrients merit to the lake. Ecotouristic management including Crane wintering as visitors' attraction is part of reasonable entire Valley management and Kinneret water quality protection. Therefore collaborative management by the farmers and tourism managers is vital. A collaborative solution between farmers, nature authorities, water managers, land owners, and regional municipalities was budgeted and implemented. Money was allocated for the renting of a 40 ha field block in the valley dedicated as "Feeding Station" where purchased Corn seeds are given to the cranes twice a day. Feeding start in late December and continue until early March when the Cranes fly back to Europe for breeding. Cranes which land prior to Mid-December are deported aimed at reducing number of potential feeders, prevention damage and reduction of the cost of Corn seeds. This achievement initiated benefits for both the landowner farmers by income resource as half a million bird visiting watchers (priced entrance) while the Hula Valley effluents were not significantly deteriorated.

It is suggested that Cranes do not contribute a significant addition of TP to lake Kinneret and the Epilimnion increase is the result of internal sources. Moreover, positive regressions were indicated between River Jordan discharge and nutrient inflow loads which is $r^2 = 0.596$, ($p < 0.0001$) for TP. Independently, the discharges in the Jordan River were declined since the mid-1980's from 15 to <10 m^3/s caused by precipitation decline.

The reconstruction of the old Hula native Flora and Fauna indicated approximately 300 bird species 12 fish species, 40 plant species observed in the Hula Valley.

The Eco-Touristic Crane Project was designed to be a part of a comprehensive objective aimed at establishment of watershed and lake Kinneret ecosystems sustainability.

The Hula Reclamation Project was aimed at ensuring sustainability of modified eco-systems by bridging over the conflict between agriculture development, Kinneret water quality protection and nature conservation. The tension between farmers, water managers, nature preservation was reduced, and collaboration came instead. The outcome of the HP was renewal of an ecosystem, which has become a tourist attraction including enriching the biological diversity.

8. Conclusive summary

The management of Lake Kinneret and its watershed require a national attempt to ensure their sustainability. These ecosystems are crucial for the nation and their protection is the national concern. Their functional efficiency can be achieved by long term managerial operation conducted by principles of sustainability.
The outcome of this paper evaluations are the following recommendations: Lake Management: (1) Shorter length of residence time to enhance water exchange and input of desalinized waters and together with pumped withdraw for supply and Dam open policy and lowering of WL are accepted options; (2) Recommended WL range between 208.8 and 213 mbsl with annual fluctuated amplitude of 1.5 m; (3) Enhance nitrogen supply to the epilimnion to encourage Peridinium bloom renovation; (4) Stocking of *Sarotherodon galilaeus* and *Hypophthalmichthys molitrix* and implementation and enforcement of fishery regulations; (5): Renewal bleaks fishery; (6): Mowing of aquatic vegetation in public beaches; Management of the Watershed: Enhance peat soil moisture through continuation of the "Peat Soil Convention"; agricultural maintenance accompanied by eco-tourism with reasonable population size of cranes and regulated number of visitors.

Author details

Moshe Gophen
MIGAL, Kiryat Shmone, Israel

*Address all correspondence to: gophen@Migal.org.il

References

[1] Gophen M, Gal I. Lake Kinneret. Ministry of Defence and Kinneret Authority; 1992. p. 335. (in Hebrew)

[2] Gophen M. Different Kinneret. Galilee Books Publisher and Migal-Scientific Research Institute; Glilit Publisher. 2019c. p. 158. (in Hebrew)

[3] Serruya C, editor. Lake Kinneret, Monographiae Biologicae (Volume 32). The Hague-Boston-London: Dr. W. Junk Publishers; 1978b. p. 501

[4] Serruya C. Chapter: Geography. In: Monographiae Biologicae Vol. 32. Rv, The Hague-Boston-London: Publisher Junk; 1978a. pp. 7-13

[5] Gophen M. Ecological Research in the Lake Kinneret and Hula Valle (Israel) Ecosystems. Scientific Research Publishing Inc; 2018c. p. 335

[6] LKDB. Kinneret Limnological Laboratory, IOLR: Lake Kinneret Data Base and annual reports. Mekorot Water Supply Co. Northern Region Monitor Unit 1970-2018. Data Base, Annual Reports. Jordan River Discharges at Gesher Huri (Gesher Ha'Pkak); 1969-2000

[7] Gophen M. The Hula Valley (Israel): From nature to anthropogenic management, a review; social sciences. Journal of Social Sciences. 2016b, 2016;**1**(7):1-10. DOI: 10.18533rss.v1i7.42

[8] Givati A, Rosenfeld D. Possible impacts of anthropogenic aerosols on water resources of the Jordan River and the sea of Galilee. Water Resources Research. 2007;**43**(2007):1-15. DOI: 10.1029/2006WR005771

[9] Givati A, Guillaume T, Rosenfeld D, Paz D. Climate change impacts on streanflow at the upper Jordan River based on an ensemble of regional climate models. Journal of Hydrology: Regional Studies. 2019;**21**:92-109. DOI: 10.1016/j.ejrh.2018.12.004

[10] Gophen M. Hydrology and management of Lake Kinneret aimed at water quality protection. In: Water in the Middle East and in North Africa: Resources, Protection, and Management. Springer-Verlag; 2003b. pp. 41-54

[11] Gophen M. Hydrology and management of Lake Kinneret aimed at water quality protection. In: Zereiny F, Jaeschke W, editors. Water in the Middle East and in North Africa: Resources, Protection, and Management. Springer-Verlag; 2004b. pp. 207-226

[12] Zohary T, Sukenik A, Berman T, Nishri A, editors. Lake Kinneret: Ecology and Management. NY: Springer; 2015. p. 683

[13] Gophen M. Climate change and water loss in the Kinneret Drainage Basin. Land Use Policy. 2018b;**80**(2019):424-429. DOI: 10.1016/j.landusepol.2018.03.008

[14] Gophen M, Meron M, Levin-Orlov V, Tsipris Y, Peres M. Climate change, regional water balance and land use Policy, in the watershed of Lake Kinneret (Israel). The Open Ecology Journal. 2020;**10**:200-224. DOI: 10.4236/oje..2020.104014

[15] Znovar Oved Gobi Ltd, Shacham G, Tsaban H, Avnimelech Y, Ofer A. Hula Project 2nd Stage, Development Program, Chapter: Opinion about Agricultural, Water consumption, Environmental and Touristic changes in the Hula Valley. Interim Report; 2011. p. 31. (in Hebrew)

[16] Gophen M. The impact of water level decline on water quality in the Epilimnion of Lake Kinneret (Israel):

Perennial perspectives. The Open Ecology Journal. 2014;**4**:892-906. DOI: 10.4236/oje.2014.414075

[17] Gophen M. Climate and water balance changes in the Kinneret watershed: Review. Open Journal of Modern Hydrology. 2020b;**10**:21-29. DOI: 10.4236/ojmh.2020.1022002

[18] Gophen M. Long term (1970-2001) eco-hydrological processes in Lake Kinneret and its watershed. In: Zereiny F, Hotzl H, editors. Climatic Changes and Water Redsources in the Middle East and North Africa. Springer-Verlag; 2004a. pp. 373-401

[19] Gophen M. Long term (1970 – 2001) eco – Hydrological processes in Lake Kinneret and its watershed. In: Zereini H, editor. Climatic Changes and Water Resources in the Middle East and in North Africa, Invited Chapter. Springer Verlag; 2008. pp. 373-402

[20] Hazan D, Stein M, Agnon A, Marco S, Nadel D, Negendank JFW, et al. The late pleistocene-holocene limnological history of Lake Kinneret (sea of Galilee). Quaternary Research. 2005;**63**:60-77

[21] Vossel H, Roeser P, Litt T, Reed JM. Lake Kinneret Israel: New insight into Holocene regional palaeclimate variability based on high –resolution multi-proxy analysis. The Holocene. 2018;**2018**:1-16. DOI: 10.1177/0959683618777071

[22] Gophen M. Water quality management in Lake Kinneret (Israel): Hydrological and food web perspectives. Journal of Limnology. 2003a;**62** (Suppl. 1):91-101

[23] Gophen M. Relation significance between hydrological residence time and phytoplankton dynamics in Lake Kinneret (Israel). The Open Ecology Journal. 2019b;**9**:479-492. DOI: 10.4236/oje.2019.911031

[24] Gophen M, Smith VH, Nishri A, Threlkeld ST. Nitrogen deficiency, phosphorus sufficiency, and the invasion of Lake Kinneret, Israel, by N2-fixing cyanobacterium Aphanizomenon ovalisporum. Aquatic Sciences. 1999;**1**:1-14

[25] Gophen M. Ecohydrological management of Lake Kinneret: A case study. Ecohydrology and Hydrobiology. 2004c;**4**(4):397-408

[26] Gophen M. The impact of available nitrogen deficiency on long term changes in the Lake Kinneret ecosystem. Open Journal of Modern Hydrology. 2015;**2015**(5):147-157

[27] Gophen M. The impact of nitrogen and phosphorus dynamics on the Kinneret phytoplankton: I: Cyanophytes-Peridinium alternate. Open Journal of Modern Hydrology. 2017;**7**:257-273. DOI: 10.4236/ojmh.2017.74015

[28] Gophen M. The replacement of peridinium by cyanobacteria in Lake Kinneret (Israel): A commentary review. Open Journal of Modern Hydrology. 2019a;**2019**(9):161-177. DOI: 10.4236/ojmh.2019.94009

[29] Nishri A, Brenner IB, Hall GEM, Taylor HE. Temporal variation in dissolved selenium in Lake Kinneret (Israel). Aquatic Sciences. 1999;**61**(3):215-233

[30] Nishri A, Sukenik A. Monitoring of Selenium Species in Lake Kinneret and its drainage basin. IOLR Report # T23/2012; 2012. (in Hebrew)

[31] Nishri A. Chapter 19.4.3. Atmospheric sources: Dry deposition (dust). In: Zohary T, Sukenik A, Berman T, Nishri A, editors. Lake Kinneret: Ecology and Management. Springer Verlag; 2014. pp. 341-346

[32] Smith VH. Low nitrogen to phosphorus ratios favor dominance by blue-gree algae in lake phytoplankton. Science. 1983;**221**:669-671

[33] Nishri A, Gavrieli I. Integrative Approach to the study of the impact of Water Bodies in the Hula Valley on Lake Kinneret. 2nd Annual Report submitted to the Water Authority, IOLR Report No. T8-2019; 2019. p. 22. (in Hebrew)

[34] Gophen M. A review on modeling of Kinneret salinity with practical recommendations. Open Journal of Modern Hydrology. 2016a;**6**(3):129-139

[35] Gophen M. Lake Kinneret (Israel) dam open management dilemma: A commentary review. The Journal of Japanese Studies. 2020a;**1**(2):1-5

[36] Gophen M, Sonin O, Lev M, Snovsky G. Regulated fishery is beneficial for the sustainability of fish population in Lake Kinneret (Israel). The Open Ecology Journal. 2015;**5**:513-527. DOI: 10.4236/oje.2015.510042

[37] Gophen M. Tilapias stock suppression by great cormorant (Phalacrocorax carbo) in Lake Kinneret, Israel. Open Journal of Modern Hydrology. 2017a;**7**:153-164. DOI: 10.4236/ojmh.2017.72009

[38] Gophen M, Sonin O, Golani D. How cormorant impact the Kinneret ecosystem (Israel)? SIL Newsletter. 2013;**63**:20-21

[39] Gophen M. Reforming provisional (2007-2008) Sarotherodon galilaeus landing decline in Lake Kinneret (Isreal). Open Journal of Modern Hydrology. 2018a;**8**:13-27. DOI: 10.4236/ojmh.2018.81002

[40] Serruya C, Gophen M, Pollingher U. Lake Kinneret: Carbon flow Patternand ecosystem management. Archiv für Hydrobiologie. 1980;**88**(3):265-302

[41] Gophen M, Crisman T, Zalidis G. Charting a course for ecosystem services in Lake Kinneret, Israel. SIL Newsletter. 2012;**61**:13-16

[42] Gophen M. Partnerships between the managements of cranes (Grus grus) and Kinneret water quality protection In the Hula Valley, Israel. Open Journal of Modern Hydrology. 2017b;**7**:200-208. DOI: 10.4236/ojmh.2017.72011

Chapter 11

Promising Water Management Strategies for Arid and Semiarid Environments

Adel Zeggaf Tahiri, G. Carmi and M. Ünlü

Abstract

Water is the most limiting factor for crop production in arid and semiarid areas. The search of promising water management strategies is foremost for achieving highly productive and sustainable agriculture. Irrigation water management, water conservation, and nonconventional water use for agriculture are key issues to be considered by the National Agricultural Research Systems (NARS) in these areas. According to climate change scenarios and population growth predictions, these countries will undergo even severe water scarcity levels. Failure of resolving food production challenge will exacerbate tensions between countries, wars, and illegal immigration and compromise human, social, economic, and sustainable development in these areas. However, the search for innovative solutions to water scarcity must comply with societal values, environmental sustainability, and market growth.

Keywords: aridity, water management, irrigation water management, water conservation, nonconventional water use, crop production, sustainability, environment

1. Introduction

Most of the Earth's water resources comprise of saline water (97.5%) covering 70% of the Earth's surface. Only 1.2% of the remaining 2.5%, which is called freshwater, is surface water and other freshwater, and it is this water which can be used for all living organisms. Therefore, renewable freshwater resources are finite and unequally distributed geographically [1]. On the other hand, the world population is growing at a rate of ~73 million per year [2], while the freshwater withdrawal, which has already tripled since 1965, is increasing at a rate of 64 km^3 $year^{-1}$ [3]. Moreover, aridity is a major economic, social, and environmental concern to the international community. It is seriously constraining the global food security, ecosecurity, socioeconomic stability, as well as sustainable development.

These will be undoubtedly the major challenges for humanity in the twenty-first century and beyond. While aridity is a natural phenomenon, humans also impact indirectly water through land use change and alterations in climate through fossil fuel combustion [4]. The desiccation of the Aral Sea which started back to the period of the Soviet Union is one of the documented examples. Severe and widespread ecological, economic, and social consequences that are progressively worsening have resulted from the Aral's recession [5].

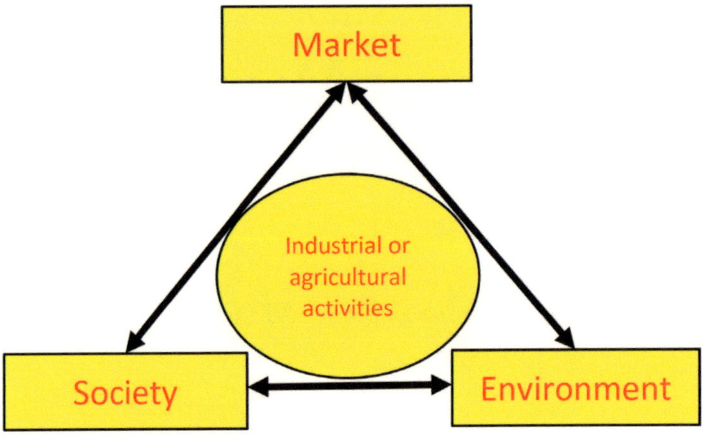

Figure 1.
A sustainable framework for industrial and agricultural activities.

To our understanding, a production model prioritizing only economic indicators such as market shares and huge benefits is one human action that exacerbates aridity, which has to be thoroughly considered. Such models could be profitable at short and medium terms but turn to the opposite at the long term since resources are undermined. In agriculture, they imply an intensive use of input resources: water, land, and plant material. This process causes water pollution, aquifer depletion, land salinization, forest clearance, etc. which leads ultimately to the habitat degradation of the Earth.

Instead, we propose the following framework as basis for any human activity (**Figure 1**). Only businesses following this pyramidal network will comply with societal values, environmental sustainability, and market growth.

In the following, promising strategies will be discussed which aim to a better resource management and hence sustainable development with the objective to reduce the negative impact of aridity on humanity future.

2. Irrigation management

Globally, water withdrawal for agriculture is estimated to 70%, 11% for domestic, and 19% for industrial uses [3]. Although there is a slight variation among North and South Mediterranean countries (**Figure 2**), depending on whether a country is heavily industrialized or not, the agricultural sector remains the largest water user for the majority. Globally, the production of irrigated crops is a predominant water "consumer" given that ~70% (~3 trillion m^3) of totally abstracted fresh hydro-resources is exploited by the agricultural sector [7]. It is obvious that any economy on agricultural water will benefit largely to other sectors.

At a global scale, agricultural water losses are enormous (**Figure 3**) reaching 55% of available irrigation water. They are caused by irrigation system, farm distribution, and field water application mismanagement. Only 45% of irrigation water is effectively used by crops. There is an urgent need to address these deficiencies and to improve water use efficiency at crop field level.

Studies showed that localized irrigation of crops is better than continuous irrigation [8, 9]. A comparison between frequent and moderate irrigation regimes for maize crop was carried by [9]. **Figure 4** shows a summary of the typical patterns of energy balances over maize field, soil surface, and maize canopy by the double layer Bowen ratio energy balance (DOLBOREB) system during both water regimes. Globally, no major differences were observed for energy balance patterns between

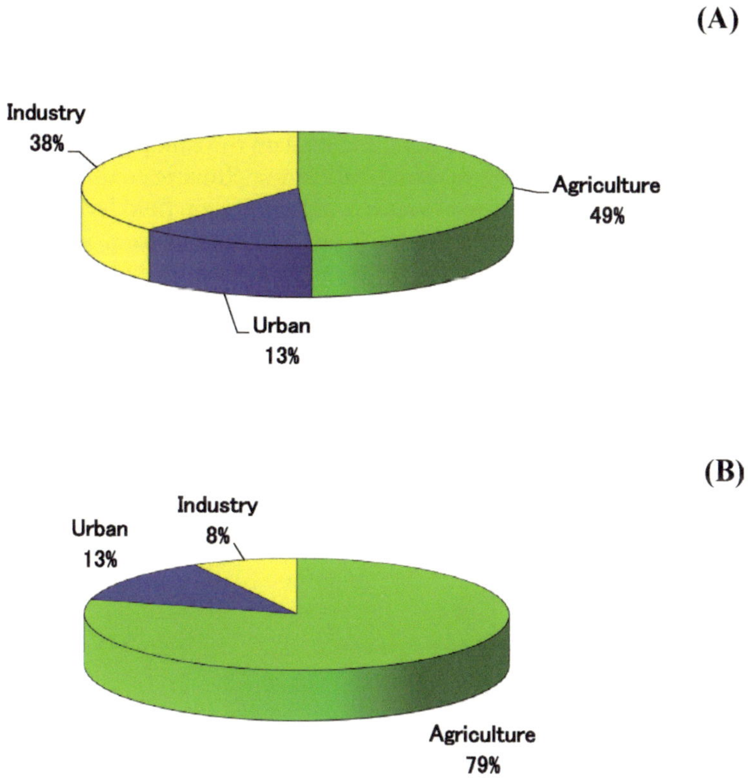

Figure 2.
Water sector allocation for North Mediterranean (A) and South Mediterranean countries (B) [6].

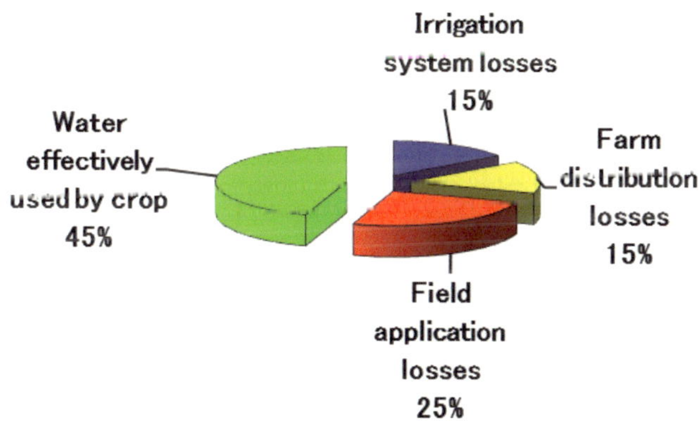

Figure 3.
Water usage by crop and losses in irrigation system [6].

the two water regimes since evapotranspiration at maize field level remained high during both regimes. In fact, maize field energy balance measurements alone provide virtually no information on how soil surface and canopy energy balances are partitioned. This shows clearly the limitations of considering crop field evapotranspiration as a whole, especially when addressing such important issues as would be water use efficiency improvement in arid areas. A number of factors have contributed to this situation. The high cost of the equipment involved in such experiments and the inherent errors associated with the use of different measurement devices and measurement scales tremendously hinted the large-scale adoption of such techniques either by research scientists and/or by irrigation practitioners [9].

The DOLBOREB system indicated that soil had a major impact on maize canopy energy balance. It also showed that a frequent irrigation regime is not necessarily a synonym of maximum plant transpiration (**Figure 4**). Ham et al. [8] also concluded that a wet soil appears to reduce crop transpiration (λE_c) by acting as a sink for advective energy while reducing the radiation load on the canopy.

Future studies for other crops and under different climatic conditions are needed to improve our knowledge of water relations at crop field level. Examining the effect of factors such canopy size, crop type, plant water stress, etc. on soil surface and canopy energy balances is of considerable importance. Energy flux data generated by the DOLBOREB system would be useful for building evapotranspiration and crop growth models. This irrigation management system would save about 30–35% of the water used at crop field level [10].

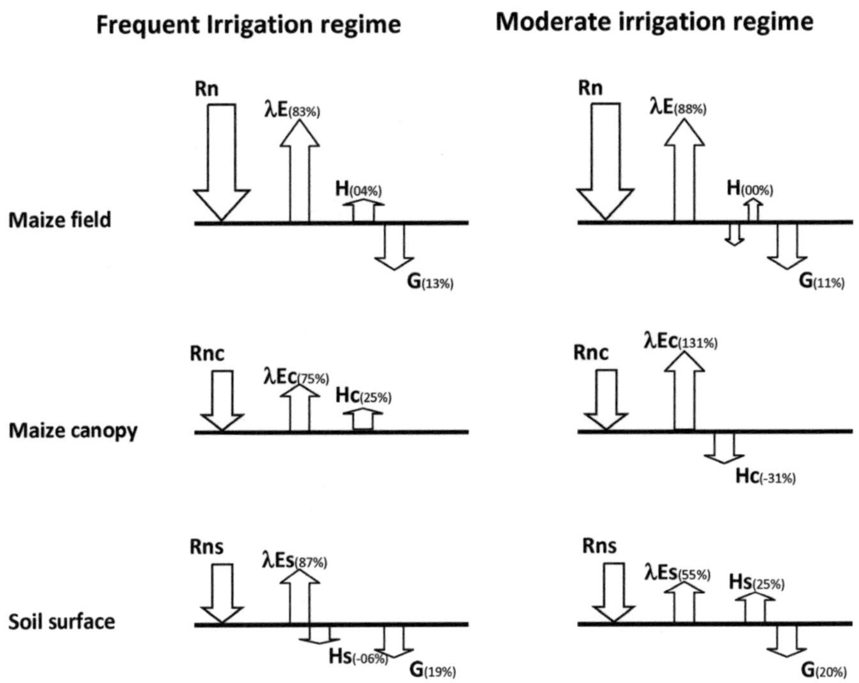

Figure 4.
A summary of the typical patterns of energy balances over maize field, soil surface, and maize canopy by the DOLBOREB system during frequent and moderate irrigation regimes.

With (R_n) net radiation; (λE) latent heat flux, (H) sensible heat flux, and (G) soil heat flux at maize field level. (R_{ns}) net radiation reaching the soil surface; (λE_s) soil surface latent heat flux, and (H_s) sensible heat flux from the soil surface. (R_{nc}) net radiation absorbed by the canopy; (λE_c) latent heat flux, and (H_c) sensible heat flux from the maize canopy.

3. Water conservation

Dryland regions occupy about 41% of the Earth's terrestrial surface and are home to more than a third of the world's population (e.g., [11]). Water is a primary limiting factor to agricultural development in these regions where the local population is suffering from food shortage. Runoff generated as a result of rainfall occurrence infiltrates to the shallow soil depth and is mostly being lost to after-rain evaporation, and the rest of the runoff is lost by strong flows to seas. Intensive agricultural practices and civil project development result in large impermeable areas, soil compaction, and crust generation that cause more runoff to be lost. Increasing runoff velocities lead to intensive erosion processes and land degradation and eventually make the region more arid. Over 17.5% of the global land area is exposed to wind and/or water erosion processes [12].

More efficient management of runoff known as runoff harvesting or runoff farming may be used for food and energy production, flood and erosion control, and landscape development [13–15]. In terms of combating desertification and land degradation, water harvesting appears to be a viable solution [13]. Runoff farming allows agricultural activity in areas that normally do not receive enough rainfall [16, 17]. This is achieved by concentrating rainfall from a collecting area (catchment) into a smaller and lower lying receiving area, where the water is stored in the soil profile, allowing its cultivation.

Hydrological aspects of these systems, especially with respect to runoff generation, have been reviewed [14, 18].

However, the use of the collected water for agricultural purposes and preventing land degradation has received little attention. The salient feature of this technique is that large amounts of water are collected a few times during the short rainy season. The collected water is ponded in walled fields and percolates to considerable depths. During the dry season, no water is added. These conditions affect plant production. To use stored water as efficiently as possible, soil evaporation and deep percolation should be minimized, and transpiration regulated to allow plants to produce biomass throughout the dry season [19].

Evaporation can be controlled by increasing tree density or mulching the soil, thus reducing the radiation that reaches the soil surface [20]. Alternatively, a similar effect may be achieved by introducing an annual intercrop. Such a crop is likely to consume water from the upper layers, part of which would otherwise evaporate directly from the soil surface. Deep-percolating water can be exploited by plant species with deep-root systems, without necessarily competing with the annual crop. The combined cultivation of shallow rooting annuals and deep-rooting perennials is proposed as a system that uses the stored water efficiently.

There are macro-catchment systems (**Figure 5**) designed to collect runoff from relatively large catchment areas used for water storage in the root zone for a group of trees or plants and micro-catchment systems designed to collect runoff from relatively small catchment areas, used for enhancing soil moisture storage in the crop rooting zone for individual crop planted in a shallow pit or micro-basin [21].

Micro-catchments for water harvesting have been tested in the Negev Desert, Israel, for decades [14]. The idea was to use runoff water for growing trees in such

Figure 5.
Flooded macro-catchment.

a) b)

Figure 6.
Traditionally designed micro-catchment system: (a) schematic of the system and (b) flooded micro-catchment.

a way that each tree had its own small catchment area, typically less than 100 m, and store it in the root zone of an adjacent infiltration basin where a tree or bush or an annual crop is grown [14, 22–24]. The system can be built on almost any slope, enabling the farmer to use large flat areas [13] that might be a significant advantage for application in the areas where collecting large amounts of runoff is not possible.

The infiltration basin is usually a shallow depression located at the low end in the immediate vicinity of the runoff generating area (**Figure 6**).

Runoff generation at micro-catchments is affected both by the total rain amounts and average rainfall intensity [18], while the relatively absolute amount of water collected at micro-catchments is low anyway. In such circumstances, the central idea behind any micro-catchment design should be enhancing infiltration and reducing evaporation of already collected water and thus improving soil moisture storage in the crop rooting zone through the dry season. The second component of the system is the water conservation efficiency at the collection plots, i.e., in the soil profile and its further availability to trees/shrubs. The deeper the harvested water moves in the soil profile, the less part of it is exposed to evaporation [15].

The size of the runoff production area directly determines the total amount of runoff water that can be stored in the pit together with soil and rainfall characteristics, topography, etc. [25]. Reported sizes of a single plot are 100–250 m^2 in Israel, 250–400 m^2 in India, and 1000 m^2 in Mali [16].

Runoff generation at micro-catchments is also affected by the rainfall characteristics. It was shown that there is a clear relationship between runoff yields and average rainfall intensity and the degree of correlation between them improves with a decrease in the length of the gap between the rainstorms [18].

The rate of water losses by evaporation is mainly affected by radiation, climate, soil texture, soil structure, soil hydraulic properties, etc. [15]. Because of relatively low absolute amount of water collected at micro-catchments, special attention should be paid to the prevention of the stored soil water from evaporation.

Long-term micro-catchment experiments carried out at Mashash runoff harvesting experimental farm of Ben Gurion University of the Negev showed that the change of collection plot design from a flat surface to a deeper and narrower pit makes the system much more effective. Being collected in the pits, water may infiltrate deeper due to repetitive concentration of relative large water amounts at the limited area and the increased waterhead. Most trees planted inside the pits showed the much higher surviving ability comparing with trees planted at the flat plots.

Infiltration and evaporation have a different pattern in the case of water collection in the pits. Soil water infiltrates through the pit bottom and the walls, where also the surface evaporation occurs through. Additionally water is lost to evaporation through the soil surface around the pit.

Deeper pits enable water to be stored in deeper soil layers around the pit, increase the distance between the stored water and soil surface, and therefore conserve more water in soil for further use by plants.

4. Seawater agriculture

The increasing deficiency of freshwater combined with the ever expanding world population will exacerbate water use pressure between the different water user sectors (urban, industrial, and agricultural). Solving this problem will undoubtedly be the twenty-first century challenge and is necessitating that marginal quality waters including saltwater and/or seawater are strategically used to meet the water shortage without any detriment to the environment and natural resources for increasing crop production worldwide.

According to the Food and Agriculture Organization (FAO) of the United Nations [26] and World Resources Institute (WRI) in collaboration with the United Nations Development Programme (UNDP), United Nations Environment Programme (UNEP), and World Bank (WB) [27], most of the West Asia and North Africa countries are expected to fall below the water scarcity level (1.000 m^3 $capita^{-1}$ $year^{-1}$) by the year 2030. The most affected countries are Kuwait, United Arab Emirates, Saudi Arabia, Yemen, and Libya where renewable water resources (RWR) per capita will fall well below 100 m^3 $capita^{-1}$ $year^{-1}$ [26–28]. Of course, reverse osmosis factories are blooming in the Middle East and North Africa, producing almost half of the 95 million m^3 day^{-1} of desalinated water for human use worldwide [29], but will not be able to meet not in the present nor in the future the growing agricultural water demand. Undoubtedly, nonconventional water use will contribute to partially alleviate water scarcity in regions where renewable water resources are extremely scarce [28].

Halophytes have demonstrated their capability to thrive under extremely saline conditions and thus are considered as one of the best germplasms for saline agriculture [30]. Few researchers have examined halophytes under special topics as sustainable cultivation, saline agriculture, and integrative anatomy [31–34]. Much practical work remains to be done, as well as developing the basic science of halophytology [35]. Apart from the cultural and sometimes the political constraints related to it, we think that there is still a big deal of scientific and technical knowledge to be studied and discovered for a better development of seawater agriculture in desert areas.

Novel approaches to mangrove planting in desert countries have been published [36, 37]. They prove establishing mangrove trees in salty coastal lands is possible providing an appropriate mineral nutrition, i.e., nitrogen, phosphorus, and iron. Based on this finding, they devised a planting method (**Figure 7**) and used mangrove nurseries. This discovery has permitted plantation of about 1 million mangrove trees, chiefly *Avicennia marina*, in the intertidal zone of the Red Sea coast of Eritrea [36]. However, this assumption has not made the unanimity among the scientific community and is contested by some other scientists [38]. Nevertheless, such forests can provide feedstuffs and serve as nurseries for fish reproduction. These important findings deserve to be considered for future mangrove plantings and/or mangrove restoration projects in Africa's desert countries.

Figure 7.
Forestation of desert area by mangrove transplants.

Figure 8.
Two-year-old mangrove forest (El Gahra, Mauritania).

Also, other projects confirmed that even with low fertilization amounts, some plant species like *Avicennia germinans*, *Nitraria retusa*, and *Sesuvium portulacastrum* can grow in extremely salty areas as well [39]. As a result, tens of thousands of mangrove trees were planted in the Mauritanian side of the Senegal River Delta and Nouakchott seaport [39]. Two years after planting, the mangrove trees reached a height of about 2 m and constitute already a source of forage foodstuff (**Figure 8**).

Thus, certain parts of the Earth's great deserts and other water-stressed areas might be converted to mangrove forests with seawater irrigation, which might be one of the possible and relatively cost-effective approaches to mitigate desertification under global warming.

5. Conclusion

Water is the most limiting factor for crop production in arid and semiarid areas. Appropriate water resource management will undoubtedly enhance crop production and accomplish sustainable development. These objectives could be achieved by adopting the following water management strategies:

- Enhancing agricultural water use efficiency by avoiding water losses at all scales, adopting efficient irrigation scheduling, and using environment adapted crops and varieties, etc.

DOI: http://dx.doi.org/10.5772/intechopen.87103

- Water conservation for better crop production
- The use of nonconventional water resources, i.e., wastewater, brackish water, and seawater along with the corresponding resistant or tolerant species to produce forage and food

Certainly, no single strategy is currently able to thrive by itself in arid environments. Each one is adapted to a physical and social environment, as well as aridity intensity. Sometimes combined water management strategies could improve crop production in water-scarce areas. Nevertheless, in these environments, the search for better water management strategies and water use habits should be a priority for both research institutions and society.

As the world population grows and climate change consequences worsen, water scarcity will intensively affect some regions more than others. North Africa and West Asia countries, among others, will be dramatically affected, as seen above. It is the responsibility of these countries to make the bulk of research in the field for no one undergoes their level of water scarcity. In this review, we showed a set of strategies, in which combination and application greatly improve plantation and water management in arid and/or desert areas. Some strategies are still not widely implemented, and others are under investigation. However, for a particular water management strategy to be successful, it should be economically viable, respectful of social values, and environmentally sustainable. The search of innovative solutions aiming for better integrated water resource management is a big challenge for National Agricultural Research Systems (NARS), the private sector, and the society as a whole.

Author details

Adel Zeggaf Tahiri[1]*, G. Carmi[2] and M. Ünlü[3]

1 Agri.Environ.Engineering SARL, Tangier, Morocco

2 French Associates Institute for Agriculture and Biotechnology of Drylands, Ben-Gurion University of the Negev, Midreshet Ben Gurion, Israel

3 Department of Agricultural Structures and Irrigation, Faculty of Agriculture, Çukurova University, Adana, Turkey

*Address all correspondence to: zeggaf.adel@yahoo.co.uk

References

[1] Rattan L. World water resources and achieving water security. Agronomy Journal. 2015;**107**(4):1526-1532

[2] UNU. "Water Security": Experts Propose a UN Definition on Which Much Depends. Dresden, Germany: United Nations Univ; 2013

[3] FAO (Food and Agriculture Organization of the United Nations). AQUASTAT. Rome, Italy: FAO; 2015

[4] Rost S, Gerten D, Bondeau A, Lucht W, Rohwer J, Schaphoff S. Agricultural green and blue water consumption and its influence on the global water system. Water Resources Research. 2008;**44**:17. DOI: 10.1029/2007WR006331

[5] Micklin PP. Desiccation of the Aral Sea: A water management disaster in the Soviet Union. Science. 1988;**241**:1170-1176

[6] Hamdy A, Lacirignola C. Mediterranean Water Resources: Major Challenges Towards the 21st Century. Bari, Italy: IAM Editions; 1999. p. 570

[7] Ondrasek G. Water scarcity and water stress in agriculture. In: Ahmad P, Wani MR, editors. Physiological Mechanism and Adaptation Strategies in Plants Under Changing Environments I. New York, Dordrecht, Heidelberg, London: Springer; 2014. pp. 75-96

[8] Ham JM, Heilman JL, Lascano RJ. Soil and canopy energy balances of a row crop at partial cover. Agronomy Journal. 1991;**83**:744-753

[9] Zeggaf TA, Takeuchi S, Dehghanisanij H, Anyoji H, Yano T. A Bowen ratio technique for partitioning energy fluxes between maize transpiration and soil surface evaporation. Agronomy Journal. 2008;**100**:988-996

[10] Zeggaf TA, Anyoji H, Takeuchi S, Yano T. Partitioning energy fluxes between canopy and soil surface under sparse maize during wet and dry periods. In: Water Saving in Mediterranean Agriculture and Future Research Needs; 14-17 February; Valenzano, Italy: Options Méditerranéennes Série B; 56. Volume I, 2007. pp. 201-211

[11] Mortimore M. Dryland Opportunities. Gland, Switzerland, HED, London, UK and UNDP, New York, USA: IUCN; 2009

[12] Pimentel D, Harvey C, Resosudarmo P, Sinclair K, Kurz D, McNair M, et al. Environmental and economic costs of soil erosion and conservation benefits. Science. 1995;**267**:1117-1123

[13] Ben-Asher J, Berliner PR. Runoff irrigation. Management of Water Use in Agricultural Sciences. 1994;**22**:126-154

[14] Boers TM, Ben-Asher J. A review of rainwater harvesting. Agricultural Water Management. 1982;**5**:145-158

[15] Zhang S, Carmi G, Berliner P. Efficiency of rainwater harvesting of micro-catchments and the role of their design. Journal of Arid Environments. 2013;**95**:22-29

[16] Evenari M, Shanan L, Tadmor NH. Runoff farming in desert. I. Experimental layout. Agronomy Journal. 1986;**1**:29-38

[17] Hilary FR, Frasier G. Micro catchment water harvesting for agricultural production. Rangelands. 1995;**17**(3):72-78

[18] Carmi G, Berliner P. The effect of soil crust on the generation of runoff on small plots in an arid environment. Catena. 2008;**74**(1):37-42

[19] Rono J, Koech E, Kireger E, Mburu F. Effect of micro-catchments rain water harvesting on survival and growth of multipurpose trees and shrubs in Nyando District, western Kenya. International Journal of Agronomy and Agricultural Research. 2013;**3**:17-25

[20] Dahiya R, Ingwersen J, Streck T. The effect of mulching and tillage on water and temperature regimes of a loess soil: Experimental findings and modeling. Soil and Tillage Research. 2007;**96**:52-63

[21] Finkel HJ, Finkel ML. Engineering measures: Water harvesting. In: Finkel HJ, Finkel M, et al., editors. Semi-Arid Soil and Water Conservation. Florida, USA: CRC Press; 1987. pp. 93-101

[22] Mzirai OB, Tumbo SD. Micro-catchment rainwater harvesting systems: Challenges and opportunities to access runoff. Journal of Plant Science. 2010;7(2):789-800

[23] Prinz D. Water harvesting for afforestation in dry areas. In: Proceedings, 10th International Conference on Rainwater Catchment Systems; 10-14 September; Mannheim: 2001. pp: 195-198

[24] Senkondo E, Msanngi MM, Xavery ASK, Lazaro P, Hatibu N. Profitability of rain water harvesting for agricultural production in selected semi-arid areas of Tanzania. Journal of Applied Irrigation Science. 2004;**39**(1):65-81

[25] Yazar A, Kuzucu M, Celik I, Sezen SM, Jacobsen SE. Water harvesting for improved water productivity in dry environment of the Mediterranean region case study: Pistachio in Turkey. Journal of Agronomy and Crop Science. 2014;**200(5)**:361-370

[26] FAO (Food and Agriculture Organization of the United Nations). Aquastat Information System on Water and Agriculture. Rome, Italy: FAO; 2005

[27] WRI (World Resources Institute) in Collaboration with United Nations Development Programme (UNDP), United Nations Environment Programme (UNEP) and World Bank (WB). The Wealth of the Poor—Managing Ecosystems to Fight Poverty. Washington, DC: WRI; 2005. p. 254

[28] Qadir M, Sharma BR, Bruggeman A, Choukr-Allah R, Karajeh F. Non-conventional water resources and opportunities for water augmentation to achieve food security in water scarce countries. Agricultural Water Management. 2007;**87**:2-22

[29] Jones E, Qadir M, Van Vliet MTH, Smakhtin V, Kang S. The state of desalination and brine production: A global outlook. Science of the Total Environment. 2019;**657**:1343-1356

[30] Mishra A, Tanna B. Halophytes: Potential resources for salt stress tolerance genes and promoters. Frontiers in Plant Science. 2017;**8**:829

[31] Flowers TJ, Munns R, Colmer TD. Sodium chloride toxicity and the cellular basis of salt tolerance in halophytes. Annals of Botany. 2015;**115**:419-431

[32] Grigore MN, Ivanescu L, Toma C. Halophytes: An Integrative Anatomical Study. New York, NY: Springer; 2014

[33] Rozema J, Muscolo A, Flowers T. Sustainable cultivation and exploitation of halophyte crops in a salinizing world. Environmental and Experimental Botany. 2013;**92**:1-3

[34] Xu C, Tang X, Shao H, Wang H. Salinity tolerance mechanism of economic halophytes from physiological to molecular hierarchy for improving food quality. Current Genomics. 2016;**17**:207-214

[35] Yensen NP. Halophyte uses for the twenty first century. In: Weber DJ,

editor. Ecophysiology of High Salinity Tolerant Plants. Switzerland: Springer; 2006

[36] Sato G, Fisseha A, Gebrekiros S, Abdulkarim H, Negassi S, Fischer M, et al. A novel approach to growing mangroves on the coastal mud flats of Eritrea with the potential for relieving regional poverty and hunger. Wetlands. 2005;**25**(3):776-779

[37] Sato G, Negassi S, Zeggaf TA. The only elements required by plants that are deficient in seawater are nitrogen, phosphorus and iron. Cytotechnology. 2011;**63**(2):201-204

[38] Lewis RR III, Erftemeijer PLA, Hodgson AB. A novel approach to growing mangroves on the coastal mud flats of Eritrea with the potential for relieving regional poverty and hunger: Comment. Wetlands. 2006;**26**(2):637-638

[39] Auriol M, Filali MY, Zeggaf TA. Mangroves implantation in coastal arid country. In: l'Etat des Ressources en Eau au Maghreb en 2009. Rabat, Morocco: UNESCO Publication; 2010. pp. 145-162. ISBN-978-9954-8068-3-0